Accelerator Physics

Editors: F. Bonaudi
C. W. Fabjan

Springer
Berlin
Heidelberg
New York
Barcelona
Budapest
Hong Kong
London
Milan
Paris
Santa Clara
Singapore
Tokyo

Rudolf K. Bock   Angela Vasilescu

The Particle Detector BriefBook

With 8 Figures

Springer

Dr. Rudolf K. Bock
CERN, EP Division
CH-1211 Genève 23, Switzerland

Dr. Angela Vasilescu
Applied Nuclear Physics Department
Institute of Nuclear Physics and Engineering
P.O. Box MG-6
RO-76900 Bucharest, Magurele, sector 5, Romania

Editors:
Professor F. Bonaudi
Professor C. W. Fabjan
CERN, Div. PPE
CH-1211 Genève 23, Switzerland

ISBN 3-540-64120-3 Springer-Verlag Berlin Heidelberg New York

Library of Congress Cataloging-in-Publication Data.
Bock, Rudolf K., 1935- The particle detector briefbook / Rudof K. Bock, Angela Vasilescu. p. cm. ISBN 3-540-64120-3 (hardcover: alk. paper) 1. Nuclear counters–Dictionaries. 2. Nuclear counters–Handbooks, manuals, etc. I. Vasilescu, Angela, 1956- . II. Title. QC787.C6B64 1998 539.7'7–dc21
98-10327

Typesetting: Data conversion by Kurt Mattes, Heidelberg
Cover design: *design & production* GmbH, Heidelberg

SPIN 10653164 55/3144 – 5 4 3 2 1 0 – Printed on acid-free paper

This *BriefBook* briefs you: it is a brief handbook, or a much extended glossary, depending on the way you look at it. In encyclopedic format, it presents practical detectors in high-energy physics experiments, the principles, the underlying physics, and the analysis of their data, together with many references to the published literature. The resulting book is both an introduction and reference for students, scientists and engineers, or anyone dealing with experiments.

A few paragraphs of this BriefBook have been derived, with permission, from a 1984 publication by the *European Physical Society* FORMULAE AND METHODS IN EXPERIMENTAL DATA EVALUATION; we acknowledge contributions by W.W.M. Allison, C.W. Fabjan, R. Frühwirth, J. Myrheim, and M. Regler.

The book is available as an html file at:

http://www.cern.ch/Physics/ParticleDetector/BriefBook/

Book information and ordering is available at:

http://www.springer.de/phys/books/accel/accel.htm

Absorption Length. The mean free path (→) of a particle before undergoing a non-elastic interaction in a given medium. The relevant cross-section is $\sigma_{\text{tot}} - \sigma_{\text{el}}$. → also Collision Length.

Acceptance. We define the acceptance a of an experiment as the average detection efficiency (→). Frequently, the word is also used in the more restricted sense of *geometric acceptance* defined below.

Let N be the total number of events that occurred, out of which n are observed. Then the expectation values of N and n are related by

$$E(n) = a\, E(N)\,.$$

One may consider the acceptance as a function of one or more variables, or in a small region of phase space, e.g. in one bin of t for a two-body process.

By this general definition, the acceptance includes all effects that cause losses of events: the finite size of detectors, the inefficiencies of detectors and of off-line event reconstruction, dead times, effects of veto counters, etc.

Let $x = (x_1, x_2, \ldots, x_D)$ be the physical variables that describe an event, such as the momenta of the particles, positions of interaction vertices, and possibly also discrete variables like the number of particles, spin components, etc. These are random variables following a probability distribution

$$f(x)\,\mathrm{d}^D x = \frac{F(x)\,\mathrm{d}^D x}{\int_\Omega F(x)\mathrm{d}^D x}\,.$$

Ω is the allowed region for x, and the integral includes a sum over discrete variables. The non-normalized density $F(x)$ is given by the experimental conditions, i.e. beam, target, etc., and is proportional to the differential cross-section. For a sufficiently small phase space region the differential cross-section is nearly constant and hence drops out from the normalized probability density $f(x)$.

Let $\epsilon(x)$ be the total detection efficiency for an event given its physical variables x. The acceptance is then the expectation value of $\epsilon(x)$,

$$a = \int_\Omega \epsilon(x)\, f(x)\, \mathrm{d}^D x.$$

If, to a sufficiently good approximation,

$$\epsilon(x) = \epsilon_g(x)\epsilon_d$$

where $\epsilon_g(x)$ is the purely *geometric efficiency* ($\epsilon_g(x) = 1$ if the particles hit the detectors, $\epsilon_g(x) = 0$ otherwise) and ϵ_d is a constant overall detection efficiency, then

$$a = a_g \epsilon_d,$$

$$a_g = \int_\Omega \epsilon_g(x)\, f(x)\, \mathrm{d}^D x\,.$$

a_g is called the *geometric acceptance.*

Acceptances are usually estimated by Monte Carlo integration (→ [Bock98]). If one is able to simulate the experiment by generating M (pseudo-) random events $x^{(1)}, \ldots, x^{(M)}$ according to the probability distribution $f(x)\,\mathrm{d}^D x$, then the Monte Carlo estimate for a is

$$A = \frac{\sum_{i=1}^{M} \epsilon(x^{(1)})}{M},$$

with the estimated variance

$$(\Delta A)^2 = \frac{\sum_{i=1}^{M} \epsilon(x^{(1)} - A)^2}{M(M-1)}.$$

If out of M generated events m events are accepted, then for the geometric acceptance a_g one has the unbiased estimates from a binomial distribution

$$A_g = m/M\,,$$

$$(\Delta A_g)^2 = \frac{A_g(1 - A_g)}{(M-1)}.$$

If some part of the integration can be done analytically, then this will reduce the variance; fewer events are necessary, hence the computing load is reduced, sometimes substantially. We will show this by an example: assume that Ω can be subdivided into non-overlapping regions $\Omega_0, \Omega_1, \Omega_2$, that the probabilities

$$P_i = P(\Omega_i) = \int_{\Omega_i} f(x)\,\mathrm{d}^D x \quad (\text{with } P_0 + P_1 + P_2 = 1)$$

can be calculated exactly, and that the regions are chosen such that $\epsilon_g(x) = 0$ for $x \in \Omega_0$, $\epsilon_g(x) = 1$ for $x \in \Omega_1$, and $\epsilon_g(x) < 1$ for $x \in \Omega_2$; in other words, the boundary of the accepted region is contained within Ω_2. Then by restricting the generation of Monte Carlo events to the region Ω_2, one obtains the estimates

$$A_g = P_1 + \frac{m}{M} P_2$$

$$(\Delta A_g)^2 = P_2^2 \frac{m(1-(m/M)}{M(M-1)},$$

which transforms to

$$(\Delta A_g)^2 = \frac{(A_g - P_1)(1 - P_0 - A_g)}{M-1} \le \frac{A_g(1-A_g)}{M-1} .$$

If the acceptance of an experiment varies with time, then the total acceptance will be a weighted average of the acceptances in different periods of time, where the appropriate weight of a period is the number of beam particles, or the integrated luminosity. → also Cross-Section.

Aerogel Detectors. Aerogels are transparent, highly porous materials of low density, ranging from 0.05 to 0.15 g/cm^3. Silica aerogel consists of amorphous grains of silica with diameter 4–10 nm, produced by hydrolysis of Silane (Si $(OCH_3)_4$) in the presence of a solvent, which is subsequently evaporated at high temperature and under pressure, i.e. replaced by air. Typical remaining bubble sizes are of the order of 60 nm; the porosity (air content) can reach 98%.

The amount of solvent determines the final refractive index n of silica aerogel, which can be adjusted to be between that of glasses ($n > 1.2$) and that of light gases ($n < 1.002$ at atmospheric pressure and room temperature). Aerogel thus can be used for particle identification by Cherenkov radiation in the momentum domain of a few GeV/c. The use as Cherenkov counters was first demonstrated by M. Cantin [Cantin74], who observed 6 to 12 photoelectrons in a volume of 18 × 18 × 18 cm^3 for hadrons with $1 < \gamma < 5$. Today, aerogel is produced commercially, mostly as superinsulator, and has also demonstrated its practical applicability in large detectors (e.g. [Carlson86], [Poelz86]). For a reference book, → [Fricke86].

The refractive index of aerogel has been measured to obey, in rough agreement with the formula of Clausius-Mosotti, the relation

$$n - 1 = (0.210 \pm 0.001)\rho,$$

where ρ is the density of the material. The density of compact silica being about 2 g/cm^3, the ratio of pore and silica volumes can be calculated by measuring n

$$V_{\text{pores}}/V_{\text{silica}} = 0.41/(n-1) - 1 .$$

Ratios of 60 with n as small as 1.007 have been achieved. The upper limit is given by the minimum of solvent ($n = 1.06$), but by baking,

$n = 1.094$ has been reached. Typically, aerogel can be produced in blocks of 100 mm sides.

The optical quality of aerogel is determined by the homogeneity of the gel. If pores are larger than some 20% of the wavelength of radiated light, Rayleigh scattering will affect the light transmission. Absorption hence dominates at $\lambda < 250$ nm. At $\lambda = 400$ nm, the diffusion length is of the order of 10 mm.

Due to Rayleigh scattering, the directionality of the radiated Cherenkov light is somewhat lost, and classical mirror focusing techniques for light collection become difficult or inadequate. Light diffusion in walls surrounding the radiator can be used for light collection; an efficiency of 50% can be achieved. For more details, → [INTE94].

Albedo. Backscattering originally of light, more generally of energy; used also as a measure for reflectivity, and named thus from the white appearance of planets due to backscattered light. In the context of particle detectors, albedo arises when a particle enters a material, and is of relevance when a hadronic particle enters a calorimeter. The (small) fraction of energy in this albedo has its origin in the breakup of nuclei; the angular distribution of particles in backscattering is more or less isotropic; the amount of energy is a few percent of incident energy, up to about 1 GeV; it settles to a constant value of 150 to 200 MeV for higher-energy incident particles ([Dorenbosch87]). → also Hadronic Shower.

Attenuation. A name given to phenomena of reduction of intensity according to the law

$$\mathrm{d}I/\mathrm{d}t = -kI\,,$$

resulting in an exponential decay

$$I = I_0 \mathrm{e}^{-kt} = I_0 \mathrm{e}^{-t/\tau}\,.$$

In this equation t may be time (e.g. attenuation of a circulating beam) or length (e.g. attenuation of light in a scintillator) or any corresponding continuous variable. The *attenuation time* or *attenuation length* is given by τ, the time (length) over which the intensity is reduced by a factor e. Frequently I is a discrete variable (number of particles), and the factor $\mathrm{e}^{-t/\tau}$ is due to the exponential distribution of individual lifetimes. τ is then the expectation value of the distribution, i.e. the *mean lifetime*.

If the intensity at time zero is I_0 and τ is the lifetime or attenuation time, then the average intensity over a time Δt is given by $I_0(1 - e^{-\Delta t/\tau})\tau/\Delta t$.

Barn. The barn is the unit of cross-section (→), 1 barn = 10^{-24} cm^2.

Beta Ray. A radiated electron. The name survives mostly in *beta decay* (the weak decay $n \to pe^-\bar{\nu}$).

Bethe–Bloch Formula. Describes the energy loss (→) of a charged particles in matter.

BGO. Short for Bismuth-Germanium-Oxyde ($Bi_4Ge_3O_{12}$), a scintillator of high atomic number Z used in electromagnetic crystal calorimeters (→). BGO is characterized by fast rise time (a few nanoseconds) and short radiation length (1.11 cm). → [Lecoq92].

Bhabha Scattering. Scattering of electrons on positrons ($e^-e^+ \to e^-e^+$). For $p \gg m_e c$ one obtains in first order perturbation theory for the differential cross-section in the centre-of-mass system (c.m.s.):

$$\begin{aligned} &\mathrm{d}\sigma/\mathrm{d}\Omega(e^-e^+ \to e^-e^+) \\ &\quad = (r_e^2/2)(m_e c/p)^2[(1/4)(1+\cos^4(\theta/2))/\sin^4(\theta/2) \\ &\qquad + (1/8)(1+\cos^2\theta) - (1/2)\cos^4(\theta/2)/\sin^2(\theta/2)] \end{aligned}$$

with

$$\begin{aligned} r_e &= \text{classical electron radius} \\ m_e &= \text{rest mass of electron} \\ p &= \text{momentum in the c.m.s.} \end{aligned}$$

Bhabha scattering is used in e^+e^- colliders for monitoring and measuring the luminosity.

Birks' Law. Describes the light output of (organic) scintillators:

$$\Delta L \propto \frac{\Delta E}{1 + k_B(\mathrm{d}E/\mathrm{d}x)} .$$

The constant k_B depends on the particle type, and is of the order of 0.01 g/MeV cm^2. For a discussion → [Birks64].

Bjorken x. Scaling variable as used in deep inelastic scattering (→ Deep Inelastic Scattering Variables). It gives the momentum fraction carried by an inclusively observed particle. Structure functions (→) are mainly dependent on this variable.

Breit Frame. → Deep Inelastic Scattering Variables

Breit–Wigner Distribution. Probability density functions of the general form

$$P(x) = \frac{1}{\pi(1+x^2)}$$

are also known in statistics as Cauchy distributions. The Breit–Wigner (also known as the Lorentz) distribution is a generalized form originally introduced [Breit36] to describe the cross-section of resonant nuclear scattering in the form

$$\sigma(E) = \frac{\Gamma}{2\pi\,[(E-E_0)^2 + (\Gamma/2)^2]}$$

which had been derived from the transition probability of a resonant state with known lifetime ([Bohr69], [Breit59], [Fermi51], [Paul69]). In this form, the integral over all energies is 1. The variance and higher moments of the Breit–Wigner distribution are infinite. The distribution is fully defined by E_0, the position of its maximum (about

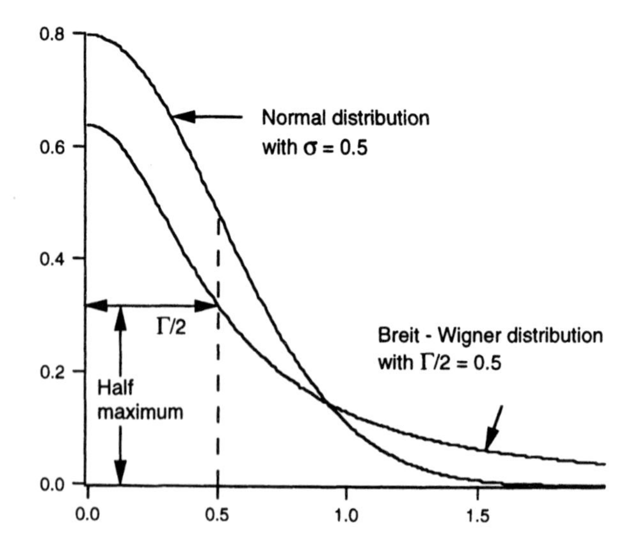

which the distribution is symmetric), and by Γ, the full width at half maximum (FWHM), as obviously

$$\sigma(E = E_0) = 2\sigma(E = E_0 \pm \Gamma/2)$$

In the above form, the Breit–Wigner distribution has also been widely used for describing the non-interfering production cross-section of particle resonant states, the parameters E_0 (= mass of resonance) and Γ (= width of resonance) being determined from the observed data. Observed Breit–Wigner distributions usually have a width larger than Γ, being a convolution with a resolution function due to measurement uncertainties.

The Gaussian curve decreases much faster than the Breit–Wigner curve in the tails. For a Gaussian, FWHM $= 2.355\sigma$, σ here being the standard deviation. In the diagram below, both curves are normalized to the same integral; a normal distribution with the same FWHM as the Breit–Wigner distribution would be even narrower.

Bremsstrahlung. Radiation emitted by a charged particle under acceleration. In particular, the term is used for radiation caused by decelerations (the word is German for braking radiation) when passing through the field of atomic nuclei *(external bremsstrahlung)*. Radiation emitted by a charged particle moving in a magnetic field is called synchrotron radiation (→).

The energy emitted by an accelerated particle is proportional to $1/m^2$, with m the rest mass of the particle; bremsstrahlung therefore plays a particularly important role for light particles; up to energies of 100 GeV, bremsstrahlung contributes substantially to energy loss in matter only for electrons. At the *critical energy* E_c, for electrons approximately given by $E_c \approx 500$ MeV$/Z$, the average energy loss by radiation and by ionization is the same (Z is the atomic number of the traversed material).

The energy spectrum of γ rays due to bremsstrahlung of electrons decelerated in the field of atomic nuclei depends on the energy levels of the atomic electrons, due to the screening effect they have on the moving particle, and on the particle velocity. The spectrum extends up to quanta of the energy of the moving particle. In the high-energy limit the probability density is given by

$$\Phi(E,k)\,\mathrm{d}k\mathrm{d}x = (\mathrm{d}x/X_0)(\mathrm{d}k/k)F(E,k)\,,$$

where k = radiated energy, x = path length, X_0 = radiation length, and F is a slowly varying function not very different from unity, that can be approximated by

$$F(E,k) = 1 - (2/3)R + R^2$$

with $R = 1 - k/E$ [Lohrmann81]. To a reasonable approximation, the amount of energy radiated per energy interval is constant.

Integration of the above formula results in the *average energy loss* per unit length which comes out to be

$$\mathrm{d}E/\mathrm{d}x = -\int k\Phi(E,k)\,\mathrm{d}k \mathrel{\widehat{=}} -E/X_0 \,.$$

(this is more or less the definition of the radiation length X_0).

In the relativistic limit, the radiated energy is contained in a narrow cone of average half-angle

$$\langle\theta^2\rangle^{1/2} = 1/\gamma = m_e c^2/E \,,$$

independent of radiated energy. For more details, → [Jauch80], [Rossi65].

The term *internal bremsstrahlung* is used to describe the radiation of non-virtual quanta, i.e. photons or gluons, by particles participating in an interaction. The formulae given for internal bremsstrahlung in electron scattering in the relativistic limit are [Bjorken64]

$$\sigma_{\text{brems}} = 2\alpha\sigma_{\text{elastic}} \log(k_{\max}/k_{\min})\{\log(|q^2|/m^2) - 1\}/\pi \,,$$

$$\Phi(E,k)\mathrm{d}k = (\mathrm{d}k/k)\{2\alpha(\log(q^2/m^2) - 1)/\pi\}F(E,k) \,,$$

where $\alpha = 1/137, q^2$ is the square of the four-momentum transfer, m the particle mass and $F(E,k)$ has been given above.

In high-energy physics, bremsstrahlung has been put to use in constructing photon beams. Coherent bremsstrahlung on crystals with incident energetic electron beams has produced photon beams with energies > 200 GeV/c ([Bilokou83], [Jackson75]).

Bubble Chamber. A particle detector of major importance during the initial years of high-energy physics. The bubble chamber has produced a wealth of physics from about 1955 well into the 1970s. It is based on the principle of bubble formation in a liquid heated above its boiling point, which is then suddenly expanded, starting boiling where passing charged particles have ionized the atoms of the liquid. The technique was perfected to work with high precision in large volumes of different liquids, embedded in a magnetic field. As liquids, they used many varieties, from the simplest nuclei, free of Fermi motion (H_2) to low-interaction length "heavy liquids" like propane (C_3H_8) or freons (Dupont's trade mark for fluor compounds, e.g. CF_2Cl_2 or CF_3Br). The liquid in a bubble chamber served simultaneously as target and as detector with a 4π solid angle coverage; stereo cameras recorded data on film. The technique was honoured by the Nobel prize award to D. Glaser in 1960. For details, → [Shutt67].

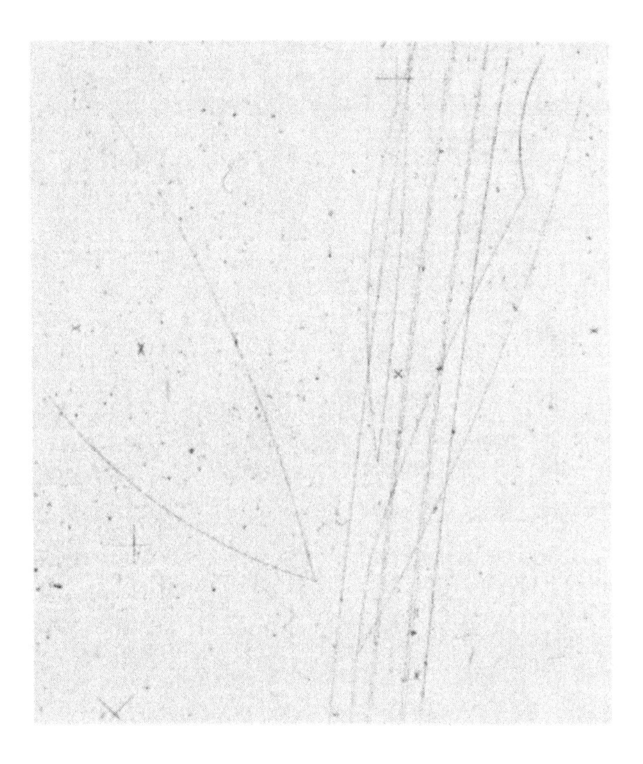

Even today, bubble chamber photographs provide the aesthetically most appealing visualization of subnuclear collisions; the above figure (from [Harigel94]) shows a historical event: one of the eight beam particles (K^- at 4.2 GeV/c) which are seen entering the chamber, interacts with a proton, giving rise to the reaction

$$K^- p \longrightarrow \Omega^- K^+ K^0 ,$$

followed by the decays

$$K^0 \longrightarrow \pi^+ \pi^-$$

and

$$\Omega^- \longrightarrow \Lambda^0 K^-, \qquad \Lambda^0 \longrightarrow p\,\pi^-, \qquad K^- \longrightarrow \pi^- \pi^0 .$$

Some chambers were built with an embedded track-sensitive target of a different (heavier) liquid; others were operated at a high repetition rate, and used in conjunction with a spectrometer of electronic detectors coupled to some trigger logic.

Bubble chamber film was scanned and measured by humans, later often assisted by computers. The projectors for scanning and measuring required substantial investment, and the teams operating them were impressive; the sharing of tasks with film as carrier of information allowed large international collaborations to emerge, with many vital tasks decentralized.

As large-volume high-precision detectors with electronic data recording became available, and physics required ever more complex triggers, and as colliders became the high-energy accelerators of choice, retirement time arrived for bubble chambers. A complete and comprehensive obituary exists in the form of conference proceedings [Harigel94]. The contribution of the technique to physics, and its role in setting up major international collaborations and in developing large-scale analysis programs is uncontested, and also this aspect was highlighted by a Nobel prize (Alvarez 1968).

Bulk. Refers to properties connected to the volume extension or the volume itself of an object, containing substance (mass). It is used mostly in characterizing the localization of effects (defects) in semiconductor detectors. E.g. *bulk defects* are produced in the volume of the detector, as opposed to *surface defects* which are created in a shallow layer at the surface of the detector.

Calorimeter. A composite detector using total absorption of particles to measure the energy and position of incident particles or jets. In the process of absorption *showers* are generated by cascades of interactions, hence the occasionally used name *shower counter* for a calorimeter. In the course of showering, eventually, most of the incident particle energy will be converted into "heat", which explains the name calorimeter (calor = Latin for heat) for this kind of detector; of course, no temperature is measured in practical detectors, but characteristic interactions with matter (e.g. atomic excitation, ionization) are used to generate a detectable effect, via particle charges. Calorimetry is also the only practicable way to measure neutral particles among the secondaries produced in a high-energy collision.

Calorimeters are usually composed of different parts, custom-built for optimal performance on different incident particles. Each calorimeter is made of multiple individual *cells*, over whose volume the absorbed energy is integrated; cells are aligned to form *towers* typically along the direction of the incident particle. The analysis of cells and towers allows one to measure lateral and longitudinal shower profiles, hence their arrangement is optimized for this purpose, and usually changes orientation in different angular regions. Typically, incident electromagnetic particles, viz. electrons and gammas, are fully absorbed in the *electromagnetic calorimeter*, which is made of the first (for the particles) layers of a composite calorimeter; its construction takes advantage of the comparatively short and concentrated electromagnetic shower shape to measure energy and position with optimal precision for these particles (which include π^0's, decaying electromagnetically). Electromagnetic showers have a shape that fluctuates within comparatively narrow limits; its overall size scales with the radiation length (→).

Incident hadrons, on the other hand, may start their showering in the electromagnetic calorimeter, but will nearly always be absorbed fully only in later layers, i.e. in the *hadronic calorimeter*, built precisely for their containment. Hadronic showers have a widely fluctuating shape; their average extent does not scale with the calorimeter's interaction length (→), but is partly determined by the radiation length.

Discrimination, often at the trigger level, between electromagnetic and hadronic showers is a major criterion for a calorimeter; it is, therefore, important to contain electromagnetic showers over a short distance, without initiating too many hadronic showers. The critical quantity to maximize is the ratio λ/X_0, which is approximately proportional to $Z^{1.3}$ (→ [Fabjan91]); hence the use of high-Z materials like lead, tungsten, or uranium for electromagnetic calorimeters.

Calorimeters can also provide signatures for particles that are not absorbed: muons and neutrinos. Muons do not shower in matter, but their charge leaves an ionization signal, which can be identified in a calorimeter if the particle is sufficiently isolated (and the dynamic range of electronics permits), and then can be associated to a track detected in tracking devices inside the calorimeter, or/and in specific muon chambers (after passing the calorimeter). Neutrinos, on the other hand, leave no signal in a calorimeter, but their existence can sometimes be inferred from energy conservation: in a hermetically closed calorimeter, at least a single sufficiently energetic neutrino, or an unbalanced group of neutrinos, can be "observed" by forming a vector sum of all measured momenta, taking the observed energy in each calorimeter cell along the direction from the interaction point to the cell. The precision of such measurements, usually limited to the transverse direction, requires minimal leakage of energy in all directions, hence a major challenge for designing a practical calorimeter.

The shower development is a statistical process (→ Electromagnetic Shower, Hadronic Shower). This explains why the relative accuracy of energy measurements in calorimeters improves with increasing energy, according to the empirical formula

$$\sigma_E/E \approx a/\sqrt{E} \oplus \beta$$

where E = energy of incident particle, σ = standard deviation of energy measurement, and a and β are constants depending on the detector type, e.g. the thickness and characteristics of active and passive layers. The overall constant β includes the systematic errors of the individual modules. Other, similar formulae, with different energy-dependent terms are in use; for more details, → Energy Resolution in Calorimeters, → Compensating Calorimeter.

From the construction point of view, one can distinguish between:

a) *Homogeneous Shower Counters.* In homogeneous calorimeters the functions of *passive* particle absorption and *active* signal generation and readout are combined in a single material. Such materials are almost exclusively used for electromagnetic calorimeters, e.g. crystals (→ Crystal Calorimeter), composite materials (like lead glass, viz. PbO and SiO_2) or, usually for low energy, liquid noble gases (→ [Walraff91], [Fabjan95a]).
b) *Heterogeneous Shower Counters (= Sampling Calorimeters).* In sampling calorimeters the functions of particle absorption and active signal readout are separated. This allows optimal choice of absorber materials and a certain freedom in signal treatment. Heterogeneous calorimeters are mostly built as sandwich counters, sheets of heavy-material absorber (e.g. lead, iron, uranium) alternating with layers of active material (e.g. liquid or solid scintillators, or proportional counters). Only the fraction of the shower energy absorbed in the active material is measured.

Hadron calorimeters, needing considerable depth and width to create and absorb the shower, are necessarily of the sampling calorimeter type.

In practical constructions the ratio of energy loss in the passive and active material is rather large, typically of the order of 10. Although performance does not strongly depend on the orientation of active and passive material, their relative thickness must not vary too much, to ensure an energy resolution independent of direction and position of showers. Only a few percent of the energy lost in the active layers is converted into detectable signal. For a discussion, → [Fabjan91].

Calorimetry is the art of compromising between conflicting requirements; the principal requirements are usually formulated in terms of resolution in energy, spatial coordinates, and time, in triggering capabilities, in radiation hardness of the materials used, and in electronics parameters like dynamic range, and signal extraction (for high-frequency colliders). In nearly all cases, cost is the most critical limiting parameter. Depending on the physics goals, the energy range that has to be considered, the accelerator characteristics, etc., some goals will be favoured over others. The span of possible solutions for calorimeters is much wider than for tracking devices, and quite ingenious solutions have been found by imaginative experimental teams over the last 15 years, since calorimeters became key components of particle detectors.

For further reading, → e.g. [Fabjan91], [Wigmans91a], [Cushman92], [Gratta94], [Gordon95], [Colas95], [Weber95], and references given there.

Cathode Strips. In multiwire proportional or drift chambers, the cathode may serve a purpose beyond supplying the electric field to make the electrons drift to the sense wires: if the cathode is made of strips with their orientation perpendicular to the anode wires they give information about the second coordinate. This technique was first used by G. Charpak and F. Sauli [Charpak73]. The avalanche on the anode wire induces a signal on several cathode strips typically an order of magnitude smaller than the anode signal, depending on the anode-cathode distance. From the pulse-height distribution on the strips, using centroid finding methods, one can determine the position of the avalanche along the anode-wire with a precision of the order of 0.1 mm with strips as wide as 5 mm at an anode-cathode distance of 3.5 mm. References are [DeWinter89], [Piuz82], [Behrend81], or [Bridges81]. The centroid finding method is specifically discussed in [Radeka80].

CCD. Short for *Charge-Coupled Device* (→).

Centre-of-Mass Transformation. A Lorentz transformation (→) from the current frame of reference to the frame of reference where a certain particle or group of particles has zero total momentum. As implied by the term "centre-of-mass", it is not possible to transform to the rest frame of a massless particle.

Charge-Coupled Device. Two-dimensional silicon device (usually abbreviated CCD) of very small ($\approx 20 \times 20\ \mu m^2$) pixels, used in commerical TV cameras and many other applications. CCDs were introduced as vertex detectors during the 1980s [Damerell81] and have had good success at e^+e^- colliders (→ [Damerell94], [McKemey96]). Individual sensors are typically of 9×13 mm size, with 400×600 pixels. They are lined up into "ladders" which in turn are arranged into cylinders. They have excellent position resolution ($\pm 5\ \mu m$), and hence are ideally suited for detecting secondary vertices from heavy-flavour events (b and d quarks with lifetimes of several times 10^{-13} seconds). Some problems with CCDs are in their noise level and in the readout speed (→ [Bross82]). The signal-to-noise ratio needs special attention due to the thin ($\approx 35\ \mu m$) active Si layer, hence is correlated with the good resolution. It is usually improved by running at low temperatures, which has the added benefit of making the device more radiation hard. The readout is serial, and hence (despite a clock frequency of 10 typically MHz or more) a total readout time of milliseconds cannot be avoided.

CCD sensors are still in a phase of research and development; some future directions are discussed in [Tsukamoto96].

Charge Division. The charge division method is used in wire chambers to measure the coordinate along the sense wire. When an avalanche occurs on a resistive sense wire of length L at a distance $x < L/2$ from the the right-end side, then the left-end-side is at $L-x$, and the signal sees less resistance towards the right than to the left: a higher signal will arrive on the right-end side. The signal ratio is given by

$$\frac{Q_{\rm L}}{Q_{\rm R}} = \frac{\beta + x/L}{\beta + 1 - x/L} ,$$

where β is the input impedance of the readout circuit divided by the wire resistance. For $\beta = 0$ this results in the simplistic non-attenuated formula

$$\frac{Q_{\rm L}}{Q_{\rm R}} = \frac{x}{L - x} .$$

Relevant studies for wire chambers can be found in [Fanet91], [Biagi86], [Dulinski83], and [Barbarino79].

Charge division in silicon strip detectors is employed not along the strips, but by grouping strips for reasons of economy (of space and electronics). As suggested in [England81], only every nth strip is connected to electronics, and interstrip capacitors couple the remaining strips. Although this gives less information on $\mathrm{d}E/\mathrm{d}x$, and inferior two-track resolution [Klanner85], the improvement in single track resolution can be substantial [Dabrowski96].

Cherenkov Counter. Detectors for charged particles using the light emitted by Cherenkov radiation (→) to measure the particle velocity β. Combined with knowledge of the particle momentum, β determines its mass. Cherenkov counters are therefore most commonly used as detectors for identifying particles, in conjunction with momentum measurements, e.g. in a tracking chamber (→ [Kleinknecht82]). Their index of refraction is carefully optimized for the particle masses and momentum range of the experiment in question.

Classification:

a) *Threshold counters* record all light produced, thus providing a signal whenever β is above the threshold $\beta_t = 1/n$.
b) *Differential counters* accept light only in a narrow range of angles ($\delta \pm \Delta\delta$) i.e. in a narrow velocity interval. Resolutions of $\Delta\beta/\beta = 10^{-5}$ have been reached. As chromatic dispersion ($n = n(\delta)$) is the major source of error at high momenta, special achromatic counters, called DISC (= directional isochronous self collimating) counters have been developed, which reach $\Delta\beta/\beta = 10^{-6}$ to 10^{-7}. Differential Cherenkov counters suffer from the low acceptance both in angle and β.
c) *Ring imaging Cherenkov counters (RICH)*: In these detectors, particles pass through a radiator, and the radiated photons are usually focused onto a position-sensitive photon detector by a fo-

cusing device (mirror). The velocity β is determined by a measurement of the radius r of the ring, on which the photons are detected. For more details, → Ring Imaging Cherenkov Counter.

Cherenkov Radiation. Cherenkov (sometimes spelled *Čerenkov*) radiation is emitted whenever charged particles pass through matter with a velocity v exceeding the velocity of light in the medium,

$$v > v_t = c/n ,$$

with

$$\begin{aligned} n &= \text{refractive index of the medium} \\ c &= \text{velocity of light in vacuum} \\ v_t &= \text{threshold velocity} . \end{aligned}$$

The charged particles polarize the molecules, which then turn back rapidly to their ground state, emitting prompt radiation. The emitted light forms a coherent wavefront if $v > v_t$; Cherenkov light is emitted under a constant *Cherenkov angle* δ with the particle trajectory, given by

$$\cos\delta = v_t/v = c/(vn) = 1/(\beta n) .$$

The maximum emission angle is given by

$$\cos(\delta_{\max}) = 1/n \quad (\text{for } v = c) ,$$

and for the threshold

$$\begin{aligned} \beta_t &= 1/n = v_t/c \\ \gamma_t &= n/\sqrt{(n^2 - 1)} . \end{aligned}$$

A more detailed treatment is given in [Allison91] and [Ypsilantis94].

The major problem of Cherenkov radiation is the modest light output: the energy loss due to ionization or excitation is two to three orders of magnitude higher than the energy lost in radiating Cherenkov light, in the energy range where photomultipliers can be used (a few eV, or about 400 nm wavelength). By its directionality, Cherenkov light can, however, be separated from the background. The useful photon yield is obtained by integrating over the range of sensitive wavelengths:

$$\mathrm{d}N/\mathrm{d}l = 2\pi\alpha[1 - 1/(n^2\beta^2)] \int \mathrm{d}\lambda/\lambda^2 ,$$

where

N = number of photons,
λ = wavelength of light,
l = length of traversed radiator,
α = fine structure constant (1/137).

For a detailed calculation of the number of photons emitted from the photocathode, the transmission factor $T(\lambda)$ and the collection factor $L(\lambda)$ have to be taken into account both for the radiator and the light guide (→), and the conversion efficiency of the photocathode must be considered (→ e.g. [Fabjan80], [Fernow86]).

Circularity. A measure of the isotropy of the distribution of particle tracks in transverse momentum p_T. It is defined to be 1 – planarity, and can be shown to be a two-dimensional equivalent of sphericity (→ Jet Variables).

Cladding. Cladding is the outer layer of a scintillator, which should ensure minimum light loss. Cladding of a scintillating fibre is usually made of thin, non-scintillating material with an index of refraction clearly higher than that of the (scintillating) core, to ensure total refraction and few light losses between the passage of the particle and the photomultiplier. For detailed discussion, → [Kazovsky96].

Cloud Chamber. Also called *Wilson chamber*, a cloud chamber is a historic device, used to make charged tracks (originally cosmic rays in pre-accelerator times) visible over a large volume. To this effect, a chamber was filled with a gas, in fact, a mixture of vapour in equilibrium with liquid, and a non-condensating gas; this mixture was brought into a supersaturated state by expansion. Condensation started around the ions generated by passing charged particles, and the resulting droplets were photographed. In a way, the cycle is just the opposite of that in a bubble chamber (→), its successor. The cycle of decompression and recompression was long, several minutes; the evaporation of droplets is slow, so they were grown to a size which made them fall to the chamber bottom by gravity. The sensitive state lasted long enough (a fraction of a second) for the chamber to be triggerable by external means (e.g. arrangements of scintillation counters).

A similar principle of using supersaturation to make visible droplets appear along particle trajectories, was used in the *diffusion chamber*; the expansion was replaced by cooling: a gas in equilibrium was continuously diffused into a cooled volume. Diffusion chambers were permanently sensitive, as the droplets moved out of the visible volume together with the gas.

Collision Length. The mean free path (→) of a particle before undergoing a nuclear reaction, for a given particle in a given medium. The collision length (also known as the *nuclear collision length*) follows from the total nuclear cross-section σ_T by

$$\lambda_T = A/(\sigma_T N_A \rho)$$

with N_A = Avogadro's number ($6.022\ 10^{23}$/mole) A = atomic weight [g/mole] and ρ = density [g/cm^3]. The probability density function for distances between successive collisions is given by

$$\Phi(x)\,\mathrm{d}x = (1/\lambda_T)\exp(-x/\lambda_T)\,\mathrm{d}x\ .$$

If one subtracts from the total cross-section the sum of elastic and quasi-elastic (diffractive) cross-sections, one obtains by the same formula the (nuclear) *interaction length* λ_1.

Some numerical values for λ_T and λ_1 are given in the following table.

Medium	λ_T [cm]	$\rho\lambda_T$ [g/cm^2]	$\rho\lambda_1$ [g/cm^2]
Fe	10.6	83.3	131.9
Al	26.1	70.6	106.4
Cu	9.6	85.6	134.9
Pb	10.2	116.2	193.7
Concrete	27.0	67.4	99.9
Scintil.	56.6	58.4	82.0

The numbers are from [Barnett96] where much more material can be found.

Compensating Calorimeter. In hadronic and combined electromagnetic/hadronic calorimeters, the energy resolution achievable for hadrons is critically dependent on the choice of absorber and active materials, and their relative thicknesses. It is important that the energy response at all energies is as independent as possible of the fluctuations in shower development, in particular the content of electromagnetic particles (electrons and gammas). This is of prime relevance for the measurement of jet energies, as in this case not only electromagnetic particles may appear during shower development, but the π^0 content (and hence the fraction of energy in the form of γ's) can be substantial in the jet before it impinges on the calorimeter.

In general, the average ratio between signals from electromagnetic and hadronic particles of the same incident energy is calorimeter- and energy-dependent, and for non-compensating calorimeters there is a higher response for electromagnetic particles, typically

$$e/h \approx 1.1\text{–}1.35\,.$$

For a compensating calorimeter, the electron/hadron signal ratio should be close to one.

Various phenomena in both active and passive layers of sampling calorimeters can be put to use to achieve $e/h \approx 1$, thus optimizing energy resolution: adjusting the relative thickness of absorber and active layers, using U^{238} as absorber for its fission capability for slow neutrons, or shielding the active layers by thin sheets of low-Z material to suppress contributions from soft photons in electromagnetic showers, are possible methods of active compensation (→ [Wigmans91a]).

The photon absorption in the (high-Z) absorber material plays a significant role, and so does the conversion of low-energy neutrons into signal, e.g. by detection of de-excitation photons; the hydrogen content in the active medium is relevant here. For a detailed discussion, → [Wigmans91b].

If high resolution is not required during readout, e.g. for triggering, corrections corresponding to compensation may also be applied by an a posteriori algorithm ("off-line"), when the shower profile (mostly the longitudinal distribution) is known (→ [Fesefeldt90a], [Andrieu93]). Just how much can be recovered by calibrations of this type, is strongly detector-dependent; [Borders94] has explored the possibilities for a specific non-compensating sampling calorimeter in detail, using individual weights for sampling layers.

Compton Scattering. Scattering of photons on free electrons ($\gamma e^- \to \gamma e^-$). Together with the photoelectric effect and pair production, Compton scattering contributes to the attenuation of γ's in matter. As the binding energy of electrons in atoms is low compared to that of passing near-relativistic particles, this is the relevant process in particle detectors. Closely related are Thomson scattering (classical treatment of photon scattering) and Rayleigh scattering (coherent scattering on atoms).

Compton scattering has a cross-section proportional to 1/E. For a discussion, → [Leo94].

Conversion Length. Used for photons, this is the attenuation length (→) due to pair production (→). The conversion length is 9/7 X_0, with X_0 the radiation length.

Coulomb Scattering. Elastic scattering of a pointlike particle with spin s on a massive point charge. Using Born's approximation, one obtains the following expression (*"Rutherford formula"*):

$$\mathrm{d}\sigma/\mathrm{d}\Omega = \frac{q^2 Q^2 F_s(\theta)}{4p^2\beta^2 \sin^4(\theta/2)}$$

with $q = ze$ the charge, p the momentum, and β the velocity of the projectile, and $Q = Ze$ the charge of the nucleus. $F_s(\theta)$ introduces an additional θ dependence coming from the spin s of the scattered particle.

Most frequently, Coulomb scattering is encountered as *multiple Coulomb scattering* and has to be integrated over many small-angle scatterings (→ Multiple Scattering).

Counter Efficiency. Probability $P(C)$ of detecting a particle traversing a detector like a scintillation or proportional counter C. $P(C)$ can be measured using a redundant setup, e.g. two detectors (T_1, T_2) of the same or smaller size and a well-defined beam such that any particle seen in T_1 and T_2 must also go through C. One obtains for $P(C)$

$$P(C) = \lim_{n(T_1 T_2)\to\infty} [n(T_1 T_2 C)/n(T_1 T_2)]$$

where $n(T_1 T_2 C)$ is the number of particles detected in T_1 and T_2 and C, and $n(T_1 T_2)$ is the number of particles seen in T_1 and T_2 regardless of what happened in C (ignoring accidental coincidences). $P(C)$ is sensitive to the operating conditions of C and to the signal processing (thresholds, electronics, dead time).

For further discussion → Detection Efficiency.

Cross-Section. The cross-section σ is a Lorentz invariant measure of the probability of interactions in a two-particle initial state. It has dimension of area (unit cm^2 or barn $\equiv 10^{-24}\mathrm{cm}^2$), and is defined such that the expected number of interactions (events) in a small volume $\mathrm{d}\boldsymbol{r}$ and a time interval $\mathrm{d}t$ is

$$\mathrm{d}N = \rho_1(\boldsymbol{r},t)\rho_2(\boldsymbol{r},t)u\,\sigma\,\mathrm{d}\boldsymbol{r}\,\mathrm{d}t = F\,\sigma\,\mathrm{d}\boldsymbol{r}\,\mathrm{d}t\,,$$

with

$$u = \sqrt{(|\boldsymbol{u}_1 - \boldsymbol{u}_2|^2 - |\boldsymbol{u}_1 \times \boldsymbol{u}_2|^2)}\,.$$

ρ_1 and ρ_2 are the number densities of the two particle species (the number of particles per volume), while $\boldsymbol{u}_1$ and $\boldsymbol{u}_2$ are their velocities. ρ and u describe the particle flux and relative direction, respectively, and can be summarily expressed by F. The cross-section can be visualized as the area presented by the *target* particle, which must be hit by the pointlike *projectile* particle for an interaction to occur.

To specify what is meant by interaction, one must specify the final state. For example, in the case of elastic scattering, if particle 1 is scattered into the solid angle $\mathrm{d}\Omega$, the cross-section for the process is denoted $\mathrm{d}\sigma$, and by definition the *differential cross-section* is $\mathrm{d}\sigma/\mathrm{d}\Omega = \frac{1}{F}\,\mathrm{d}N/\mathrm{d}\Omega$.

Cross-sections in colliding beam experiments: The beams in storage rings travel in bunches or continuously, and collide either head-on or at a small angle. The time average of the quantity $\int \rho_1\rho_2 u\,\mathrm{d}^3\boldsymbol{r}$ is called the luminosity (→) L of the collider, and describes the achieved intensity. The average event rate (counts per unit of time) is simply $L\sigma$.

Cross-sections in fixed target experiments: For N_b beam particles incident upon a fixed target the expected number of events is, if the attenuation of the beam along the target is neglected,

$$E(N_\mathrm{e}) = N_\mathrm{b}(N_\mathrm{t}/T)\sigma = N_\mathrm{b}\delta_\mathrm{t} l\sigma\,.$$

N_t is the number of target particles, T is the area of the target perpendicular to the beam direction and l is the length (thickness) of the target along the beam direction. δ_t is the number density of target particles, which is related to the mass density ρ_t by

$$\delta_\mathrm{t} = \rho_\mathrm{t}/m_\mathrm{t} = N_\mathrm{A}\rho_\mathrm{t}/A\,.$$

Here m_t is the mass of one particle, $N_\mathrm{A} = 6.022\ 10^{23}/\mathrm{mole}$ is Avogadro's number and A is the atomic weight [g/mole].

Due to attenuation of the beam, events are exponentially distributed along the target, and one way to take this effect into account is to write

$$E(N_e) = B\delta_t l\sigma$$

where

$$B = \int_0^l N_b \exp(-x/\lambda_t)\,dx = N_b(\lambda_1/l)(l - \exp(-l/\lambda_1))$$

is the effective number of beam particles and λ_1 is the interaction length (→).

The observed number of events, N_{obs}, allows one to estimate the cross-section. N_{obs} is subject to observational losses; it has expectation value and variance

$$E(N_{obs}) = \mathrm{var}(N_{obs}) = A\,B\,\delta_t\,l\,\sigma\,.$$

The factor $A \leq 1$ includes all effects that cause loss of events in an experiment; it may be called the acceptance (→), although the term "acceptance" is often used in the more restricted sense of "geometric acceptance". The exponential distribution of events along the target is only one of the many effects that must be taken into account in calculating the acceptance.

An unbiased estimator for σ is

$$\sigma_{est} = N_{obs}/(A\,B\,\delta_t l)\,,$$

with the (estimated) variance

$$\mathrm{var}(\sigma_{est}) = N_{obs}/(A\,B\,\delta_t l)^2\,.$$

If experimental conditions, like beam intensity, geometry, etc., vary with time, then the factor $A\,B$ is replaced by $\sum A_i B_i$, summed over periods in time such that conditions do not vary during one period. In this way all events are assigned equal weight, whereby $\mathrm{var}(\sigma_{est})$ is minimized.

Formulae for cross-sections of specific processes are given in [Barnett96].

Crystal Calorimeter. A calorimeter made of homogeneous cells and towers, of high-Z inorganic scintillating materials, e.g. NaI(Tl), BGO, BaF_2, CeF_3, or CsI.

The materials may be *impurity-activated* scintillators such as NaI(Tl), or *pure* crystals like BGO; the latter are free from problems associated with non-uniform dopant distribution, but are usually also harder to grow. Materials are characterized by their radiation length X_0 (ranging from 1.1 to 2.5 cm), and interaction length λ (ranging from 22 to 41 cm), the light yield in photons per MeV and wavelength of emitted light, signal rise and decay times, afterglow (duration of light emission after excitation), etc. Overall, it is the physical size (cost) and the energy resolution that usually count most. Resolutions down to $\sigma_E/E \approx 0.02/\sqrt{E} \oplus 0.005$ have been discussed ([Ferrere92], [Lecoq93]). For a comparison of crystals, → [Gratta94] or [Majewski92], also [Barnett96].

Dalitz Pair. The most frequent occurrence of Dalitz pairs is in the electromagnetic decay of the π^0, dominated by $\pi^0 \rightarrow \gamma\gamma$. In 1.19% of the decays, one of the photons never materializes, and the observable reaction is

$$\pi^0 \rightarrow e^+e^-\gamma;$$

the electron/positron pair produced in such a decay is called a Dalitz pair. Usually, the two charged particles have a wide opening angle.

Dalitz Plot. The Dalitz plot is a way to represent the entire phase space, viz. all essential kinematical variables, of any three-body final state in a scatter plot or two-dimensional histogram. Dalitz introduced it in 1953 [Dalitz53]. Let a reaction be

$$1 + 2 \rightarrow 3 + 4 + 5 \, .$$

For a given incident energy, two of the three possible two-body effective masses of the final state fully describe the reaction. Choosing as abscissa and ordinate the squares of the effective masses (p_i = four-momentum of track i)

$$m_{34}^2 = (p_3 + p_4)^2$$

$$m_{45}^2 = (p_4 + p_5)^2$$

the third effective mass squared (m_{35}^2) is constant along lines at 45°, as

$$m_{34}^2 + m_{45}^2 + m_{35}^2 = m_{12}^2 + m_3^2 + m_4^2 + m_5^2 = \text{const.}$$

For fixed p_1 and p_2, i.e. fixed total energy, the physical region of a Dalitz plot is inside a well-defined area, and in the absence of resonances or interferences can be shown to be uniformly populated. Resonant behaviour of two of the final state particles gives rise to a band of higher density, parallel to one of the coordinate axes or along a 45° line.

The following graph shows a Dalitz plot for the annihilation process

$$p + \bar{p} \rightarrow \pi^0 + \pi^0 + \pi^0,$$

an example taken from the Crystal Barrel experiment, of exceptionally high statistics (here some 750 000 events, from [Landua96]):

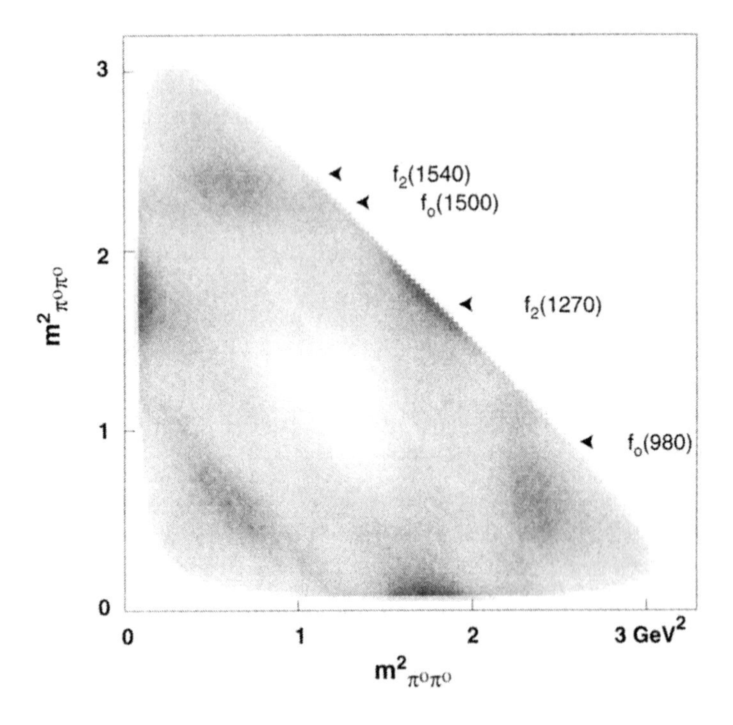

In its projection, the various f_0- and f_2-resonances are clearly visible. The plot has an inherent 6-fold symmetry, as all particles are identical.

Dark Current. A type of noise that occurs in light-sensitive detectors, most typically in photomultipliers (→). They emit a small signal even in the absence of light, mostly due to thermal activity in the photocathode and the dynodes, hence operation at low temperature can alleviate the effect. Leakage currents are sometimes also named "dark" currents, e.g. in semiconductor detectors.

Dead Time. Absolute dead time is a span of time during which a detector, or an associated readout system, is unable to record new information. Relative dead time is the average ratio of dead time to total time.

During dead time, the detector is typically busy with collecting the information generated by a previous event, and is disabled, viz. made insensitive to new events; if it remains active and continues to record information, one will have the problem of analysing combined signals (so-called *pileup*), from which information has to be extracted. For more discussion, → Detection Efficiency, Trigger Efficiency; → also [Leo94].

Deep Inelastic Scattering Variables. In the inclusive reaction

$$l + N \rightarrow l' + H$$

with l, l' leptons, N a nucleon, and H any hadronic system, the following kinematic variables are frequently used to describe the interaction (p_i is the 4-vector of particle i, and as usual $p_i \cdot p_j = E_i E_j - \mathbf{p}_i \cdot \mathbf{p}_j$):

The energy transfer or *hardness*:

$$Q^2 = -q^2 = -(p_l - p_{l'})^2 \approx 2\, E_l\, E_{l'} (1 - \cos\theta) \ ,$$

the energy of the transferred particle, in the N rest system

$$\nu = (p_l - p_{l'}) \cdot p_N / m = E_l - E_{l'} \ ,$$

the square of the mass of the hadronic system:

$$W^2 = p_H^2 = m^2 + 2m\nu - Q^2 \ ,$$

and the dimensionless scaling variables *Bjorken* x

$$x = \frac{Q^2}{2(p_l - p_{l'}) \cdot p_N} = \frac{Q^2}{2m\nu} \ ,$$

and *inelasticity*

$$y = \frac{(p_l - p_{l'}) \cdot p_N}{p_l \cdot p_N} = \nu / E_l \ ;$$

generally used is the square of the centre-of-mass energy of the $l + N$ system

$$s = (p_l + p_N)^2 \approx Q^2/(xy) = 2mE_l .$$

Frequently, the notation is shortened to $p = p_N$, $k = p_l$, $k' = p_{l'}$, which gives

$$q = k - k',$$
$$\nu = p \cdot q/m,$$
$$W^2 = (p+q)^2,$$
$$x = Q^2/(2p \cdot q),$$
$$y = p \cdot q/p \cdot k.$$

The corresponding Feynman diagram is the following:

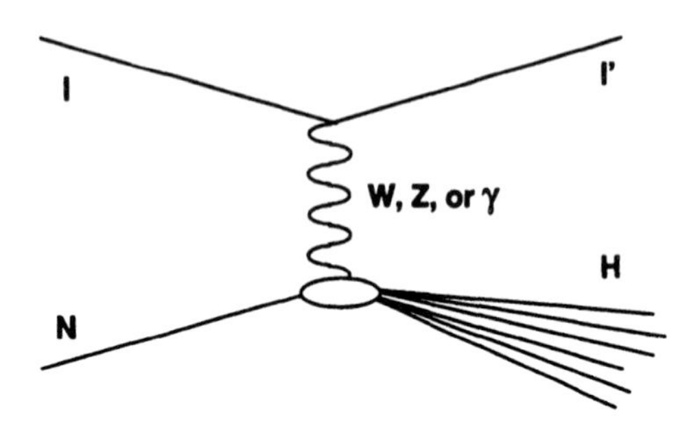

θ is the angle between the incoming l and outgoing lepton l' directions, and m is the nucleon mass. For more details, → [Abramowicz94].

Frequently, the analysis of the hadronic system is done in the *Breit frame*, defined to be the frame in which the momentum transfer q (usually the virtual photon) has no transverse component and the longitudinal (z) component is $-Q$. The plane $z = 0$ divides the event into the "current" hemisphere $z < 0$, dominated by the lepton, and the "remnant" hemisphere, opposite. For an example of analysis, → [Adloff97].

The same variables are used in *two-photon* processes, e.g. in the reaction

$$e^+ + e^- \rightarrow e^+ + e^- + X \text{ (hadronic)},$$

with one electron tagged at large angle, and the other undetected. This process is mediated by photons, one quasi-real (small four-momentum transfer k, i.e. undetected electron), the other virtual (four-momentum transfer q). The variables in this process can be viewed as deep inelastic $e\gamma$ scattering; l and l' refer to the initial and final tagged electron.

Delta Ray. Common name for → Knock-on Electron. The name dates back to the time of emulsions, when alpha and beta particles were also named.

Detection Efficiency. The probability of detecting an event if it has taken place. The event might be e.g. a collision process involving several particles, or simply the passage of one single particle through one single counter.

The detection efficiency is a function of the variables that describe the event, but depending on definition may also include the effects of other events, e.g. by *dead time* in the detector or its electronics caused by a previous event. If these variables are not completely specified, i.e. if some or all of them are random variables, then the interesting quantity is the expectation value of the detection efficiency. This expectation value is again often called the detection efficiency, although a more precise name would be *average detection efficiency.* When a cross-section (→) is to be measured, one must correct for the average detection efficiency, which is often also called the *acceptance* (→).

If the detection of the event depends upon several independent necessary conditions, then the total detection efficiency for one given event is the product of independent efficiencies. The same factorization is not necessarily valid for the corresponding average detection efficiencies, due to possible correlations.

The detection efficiency of a single counter or of a complex detector can be measured either in a test experiment or from the final data sample, if there is redundant information. To measure the efficiency of one counter, one needs a sample of events for which the detection does not depend on that counter. If in n out of N events the counter produces a signal, then

$$e = n/N$$

is an unbiased estimator for the efficiency ε, with variance

$$\mathrm{var}(e) = \varepsilon(1-\varepsilon)/N \approx e(1-e)/(N-1) \; .$$

The probability of obtaining in a measurement n events out of N, given ε, is estimated from a Poisson distribution with mean εN (→ [Bock98]). → also Counter Efficiency, Trigger Efficiency.

Diffusion Chamber. → Cloud Chamber

Diffusion in Gases. During the drift in electric fields, charged particles diffuse according to a Gaussian distribution

$$f_t(x) = \frac{1}{\sqrt{4\pi Dt}} \exp\left[-(x - tv_{\mathrm{drift}})^2/(4Dt)\right]$$

where

$$\begin{aligned} v_{\mathrm{drift}} &= \text{drift velocity} \\ t &= \text{drift time} \qquad (x_{t=0} = 0, f_0(x) = \delta(x)) \\ D &= \text{diffusion coefficient} . \end{aligned}$$

It is convenient to define a reduced drift velocity, the *mobility* at atmospheric pressure

$$\mu = v_{\mathrm{drift}}(P/E)$$

with

$$\begin{aligned} P &= \text{pressure} \\ E &= \text{electric field} \\ E/P &= \text{reduced electric field} . \end{aligned}$$

From classical arguments it can be shown that the diffusion coefficient is given by the Nernst–Einstein relation

$$D = \mu k\, T/e$$

with

$$\begin{aligned} k &= \text{Boltzmann constant} \\ T &= \text{absolute temperature} \\ e &= \text{unit charge} . \end{aligned}$$

The mobility depends on the energy distribution, the mean free path (→) and the inelasticity $\Lambda(E)$, i.e. the fraction of energy lost on each impact.

For positive ions, the following table gives some values for the mean free path λ and the diffusion coefficients D for different molecules under normal conditions (from [Schultz77] and [Sauli91]):

Gas	λ [cm]	D [cm^2/s]	μ[cm^2 sec^{-1} V^{-1}]
H_2	1.8×10^{-5}	0.34	13.0
He	2.8×10^{-5}	0.26	10.2
Ar	1.0×10^{-5}	0.04	1.7
O_2	1.0×10^{-5}	0.06	2.2
H_2O	1.0×10^{-5}	0.02	0.7

For electrons, the neutralization by ions and the attachment by molecules with electron affinity must be considered. Except for very low fields the mobility of electrons is not a constant; the mean free path varies in some gases with the electric field (Ramsauer effect), all resulting in a diffusion coefficient dependent on the electric field.

Note that the limiting accuracy is not given by the standard deviation from $f_t(x)$, but depends on the number of electrons necessary to trigger the shift-line electronics. If n electrons are produced and k electrons are needed to overcome the electronics threshold, the following formula holds:

$$\sigma_k^2 = \sigma_x^2 \sum_{i=k}^{n} (l/i^2)/(2\log(n)) .$$

For more details, → [Piuz83], [Breskin84], [Charpak84], [Peisert84], [Amendolia86], [Sauli91].

DISC. Short for directional isochronous self collimating (Cherenkov counter), → Cherenkov Counter.

Doping. Doping is the process of introducing impurities (atoms with Z different from the basic element) into scintillating materials (→ Scintillation Counter) or semiconductors (→ Semiconductor Detectors), in order to improve their detection properties.

In semiconductors, doping typically produces a region with different charge carrier concentration, or modifies other properties of the bulk (→). Electrically active dopants belong to groups III and V of the Mendeleev table. For Si, they are usually boron and phosphorus (or arsenic), to produce p- and n-type material. Typical concentrations of dopants in detector grade silicon are in the range of 10^{12} atoms/cm^3 [Dreier90].

Other dopants (e.g. metals, increased carbon or oxygen concentrations) are used for special purposes, to modify the probability of recombination of different defects and the defects kinetics, in *defect engineering*.

Bulk doping is the result of the growth process of the monocrystalline material, and depends on technology and initial material. Highly doped shallow regions, e.g. the abrupt p-n junction in a semiconductor diode detector, are produced by ion implantation, diffusion or a combination of both techniques.

A basic reference to semiconductor devices is [Sze81]; a good introductory text is [Klanner85].

Drift Angle. → Drift Velocity

Drift Chamber. A multiwire chamber in which spatial resolution is achieved by measuring the time electrons need to reach the anode wire, measured from the moment that the ionizing particle traversed the detector. This results in higher resolution and wider wire spacing than can be obtained with simple planar or cylindrical multiwire proportional chambers. Fewer channels have to be equipped with electronics, although the cost per channel is increased. Drift chambers use longer drift distances, hence are slower than multiwire chambers; therefore, they are typically not used in the primary beam, in high-rate colliders, or for triggering purposes.

Drift chambers have been built in many different forms and sizes, and they are standard tracking detectors in more or less all experiments; this is true even in high-rate colliders, where the collision rate can be shorter than the maximum drift time. Planar (or cylindrical) chambers, with the drift in the same plane as the wires, have been operated with anode wires up to 50 cm apart, but more typically distances of some 5 cm are used. Non-planar chambers, with the drift direction orthogonal to the wire plane, exist in large varieties; the most ambitious developments are large jet chambers and time projection chambers. Shortest drift times are achieved in drift tubes. Recent developments are silicon drift chambers (→) and microstrip gas chambers (→).

To translate good time resolution into spatial resolution, it is important to have a predictable electron drift velocity in the gas, and a simple relation for tracks passing under different angles; this means that the shape and constancy of the electric field needs more careful adjustment and control than in ordinary multiwire proportional chambers. In planar drift chambers, the anode wires are alternated with thick field-shaping cathode wires often called *field wires*, that reinforce the electric field right in between two anode wires. The anode wire is maintained at a positive potential, and the two field wires at the potential of the adjacent cathode wires. By choosing the proper voltages, a uniform drift field can be produced over the entire cell, for modest gaps in the order of 6 or 8 cm, and with anode wires in the order of 5 cm apart (→ Field Shaping).

Planar drift chambers measure the coordinates of the intersection of a particle track with a wire plane, by making the electrons drift in the plane. Hence multiple planes are needed to determine a trajectory; they are typically given several different wire orientations, to get different projections, thus offering the possibility of reconstruction in three dimensions.

Drift tubes, stand-alone cylindrical detectors with a single sense wire along the axis, are used in various arrangements (but typically in large numbers) when short drift time is needed, as in hadron colliders (→ Drift Tube);

Jet chambers made of multiple independent cells, with a single wire plane in a moderate drift volume, often using drift on both sides of the wire; the left-right ambiguity is resolved by staggering the wires (displacing alternate wires in the drift direction, by a small amount), so that ghost (wrongly assigned) digitizings will not result in a smooth track. The drift direction in a jet chamber is roughly perpendicular to the wire plane, with only a small amplification region. Thus a single trajectory gives rise to many hits on different wires. Given multihit electronics, two-track resolution can be very good (hence the name jet chamber); ionization sampling is also possible. The effect of a magnetic field (Lorentz angle) has to be taken into account, e.g. by tuning the potential on the cathode wires. Precision along the drift is typically ± 100 μm, and can be better than ± 50 μm with pressurized gas (jet chambers are also in use close to the vertex). Two-track resolution is 1–3 mm, and precision by charge division along the wire is a few centimetres. Fully explained examples can be found in [Blum93].

Time projection chambers or TPCs for short, share many of the properties of jet chambers; their drift volumes are larger (up to 200 cm), and the sense wires are arranged in one end face; the effect is that no left–right ambiguity can arise, and signals induced in pads or strips near the sense wire plane can be used to obtain three-dimensional information. Also, the drift direction being often along the magnetic field, diffusion is reduced. On the other hand, the long drift time and the difficulty of shaping the field are drawbacks: space charge builds up, and inhomogenities in the field can cause serious degradation of the precision. Introduction of ion-stopping grids ("gates"), careful tuning of the drift field (sometimes by an additional "potential" wire plane), and gas purity are of paramount importance to the resolution achieved in these chambers.

High-voltage plane (cathode)

Drift volume

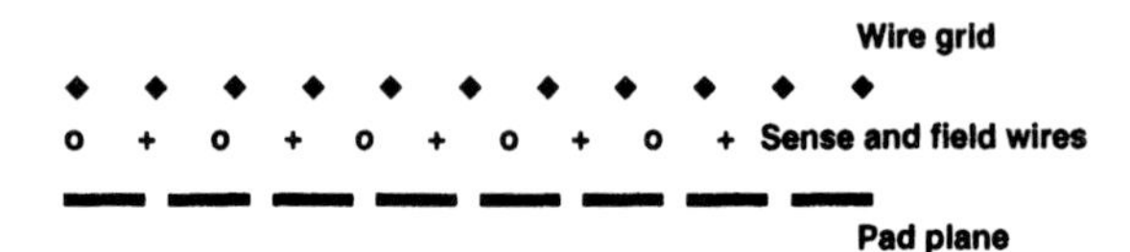

The achieved precision along the drift direction is typically ±150–200 μm, and the precision along the wire, by charge division, is a few centimetres.

For more details, → [Blum93] (with many examples), [Lohse92], [Aarnio91], and [Peisert84], also in the proceedings of various specialized conferences, e.g. [Krammer95] or [Villa86]. Basic considerations on precision can be found in [Sadoulet82], on readout of wire chambers in [Radeka91].

Drift Tube. A stand-alone coaxial cylindrical drift chamber, made of a conducting-surface cylinder acting as cathode, and a sense wire stretched in the axis of the cylinder. The function is the same as that of a *proportional tube*, with measurement of drift time added. The simple construction achieves high gain and good proportionality.

Often, tubes are made of thin metallized foils (e.g. 25 μm kapton with an evaporated conductive layer), and arranged into densely packed layers or volumes; these can be used when short drift times are at a premium, like in high-rate environments. For a tube diameter of 4 mm, the maximum drift time (at the usual drift velocity) is 40 ns; they can thus be used for triggering. Such small-diameter tubes are also called *straws*, and a collection of them a *straw chamber*. High position precision and $\mathrm{d}E/\mathrm{d}x$ measurements are difficult; mostly, if the occupancy (hit probability) is not too high, single-hit electronics will suffice.

Drift Velocity. The drift velocity v_{D} in an electric field is given by

$$v_{\mathrm{D}} = \frac{e\tau E}{2m}$$

where

$$\begin{aligned} e &= \text{charge} \\ E &= \text{electric field} \\ m &= \text{mass} \\ \tau &= \text{mean time between collisions;} \end{aligned}$$

v_{D} is typically arranged to be of the order of 50 mm/µs. In the presence of a magnetic field B, the drift velocity is reduced, and the drift direction is no longer along the electric field. The apparent drift velocity is

$$v'_{\mathrm{D}} = v_{\mathrm{D}} \cos \alpha_{\mathrm{B}} \, ,$$

where α_{B} is the drift angle given by

$$\tan \alpha_{\mathrm{B}} = 2Bv_{\mathrm{D}}/E \, .$$

This assumes B and E to be orthogonal. For details, → [Sauli91]. More general formulae are given in [Blum93].

Dynamic Range. The range of signals that can be reliably transmitted in a device, usually a digital system. Digitizing electronics are often carefully designed for the desired dynamic range to fit into the information range given by the device. Thus an analogue-to-digital converter (ADC) trying to record with acceptable resolution light pulses from minimum ionizing tracks and from energetic showers in a calorimeter, will necessarily need a large information range (usually given as a word length, e.g. 10 or more bits); often, the effective range of an ADC range is extended by giving it a non-linear response: this extends the dynamic range, preserving all relevant information, but for adding, the information has to be translated back to a linear scale. For non-linear response, the error given by the *least count* (viz. the smallest step by which digitized signals can be different) does not translate into a constant absolute error.

The dynamic range is sometimes given as the ratio between the highest and lowest signal, and may then be expressed in *decibels* (i.e. the tenfold $\log_{10}$ of this ratio).

Dynode. Part of the electron multiplication chain in a photomultiplier (→).

Effective Mass. The effective mass m_{eff} of several particles with four-momenta (→) p_i is defined by

$$m_{\text{eff}}^2 = \left(\sum p_i\right)^2 ,$$

or, in terms of energy and three-momenta,

$$m_{\text{eff}}^2 = \left(\sum E_i\right)^2 - \left(\sum \mathbf{p}_i\right)^2 .$$

In the case of two particles and expressed in scalar variables E_1, E_2, p_1, p_2 and the opening angle Θ_{12}, the expression transforms into

$$m_{12}^2 = m_1^2 + m_2^2 + 2(E_1E_2 - p_1p_2\cos\Theta_{12}).$$

The effective mass is a Lorentz-invariant variable.

Electromagnetic Calorimeter. A calorimeter (→) optimized for measuring electrons and gammas (→ Electromagnetic Shower). In general-purpose detectors, an electromagnetic calorimeter is usually followed by a hadronic calorimeter.

Electromagnetic Interactions. Interactions between electric charges, the carrier of the electromagnetic interaction being the photon.

In classical physics, electromagnetism is described using electric and magnetic fields. The basic relations between these fields and matter are expressed by Maxwell's equations (→). In quantum theory, electromagnetic interactions are described by quantum electrodynamics (QED) and can be calculated using perturbation theory (Feynman diagrams). Typical electromagnetic interactions in high-energy physics are:

- Coulomb scattering (e.g. electron–nucleon scattering),
- Bhabha scattering (electron–positron scattering),
- Möller scattering (electron–electron scattering),
- Compton scattering (photon–electron scattering),

- Bremsstrahlung (photon emission in deacceleration or acceleration),
- Annihilation (e.g. $e^+e^- \rightarrow \gamma\gamma$),
- Pair creation ($\gamma \rightarrow e^+e^-$),
- Decay of π^0.

Electromagnetic Shower. Bremsstrahlung and electron pair production are the dominant processes for high-energy electrons and photons; their cross-sections become nearly independent of energy above 1 GeV. The dominance of these electromagnetic processes and their small fluctuations distinguish the electromagnetic showers (initiated by e's and γ's) from hadronic showers (→). The π^0, decaying electromagnetically, produces two, possibly three, electromagnetic showers (→ Dalitz pair).

The cross-sections can be described in units of a scaling variable, the radiation length (→) X_0.

Secondaries produced in electromagnetic processes are again mainly e^+, e^- and γ, and most of the energy is consumed for particle production (inelasticity $\kappa \approx 1$). The cascade develops through repeated similar interactions. The shower maximum, with the largest number of particles, is reached when the average energy per particle becomes low enough to stop further multiplication. From this point the shower decays slowly through ionization losses for e^-, or by Compton scattering for photons. This change is characterized by the critical energy ε in the absorber material. ε is the electron energy for which energy loss by radiation equals the collision and ionization losses, and is approximately 550 MeV/Z. Nuclear interactions (photonuclear effects) play a negligible role.

The electromagnetic shower shape, to a good approximation, scales longitudinally with the radiation length, and laterally with the Moliere radius (→). Experimental results on shower shape have been parameterized in the following way (→ [Fabjan82]):

Shower maximum:

$$t_{\max} \approx \log(E/\varepsilon) - a \qquad [\text{in units of} X_0]$$

with

$$\begin{aligned} a &= 1.0 \qquad \text{for } e^+, e^- \\ a &= 0.5 \qquad \text{for } \gamma \\ E &= \text{energy of incident particle} \\ \varepsilon &= \text{critical energy} \ . \end{aligned}$$

Shower depth for 95% longitudinal containment:

$$t_{95\%} \approx t_{\max} + 0.08\,Z + 9.6\ [X_0]\ .$$

Transverse shower dimension (95% radial containment):

$$R_{95\%} \approx 14\ A/Z\ [\mathrm{g\,cm^{-2}}]$$

For the average differential longitudinal energy deposit over the volume of the cascade a reasonable longitudinal parametric approximation is given by:

$$DE = k\ t^{(a-1)}\ \mathrm{e}^{-bt}\,\mathrm{d}t$$

with a and b fitted from Monte Carlo or experimental data (→ [Bock81] or [Longo75]) and

t = depth starting from shower origin in units of X_0
k = normalization factor ($= E\ b^a/\Gamma(a)$).

For the average lateral electromagnetic shower development, double exponentials and Breit–Wigner distributions have been shown to fit experimental data (→ [Acosto92]).

Electron Avalanche. → Gaseous Detectors, Operational Modes

Electron Volt. The unit of energy used in high-energy physics. The eV is defined as the kinetic energy picked up by an electron when passing through a potential difference of one volt (→ Units).

Energy Flow. → Momentum Flow

Energy Loss. A particle passing through matter interacts with electrons and with nuclei, possibly also with the medium as a whole (Cherenkov radiation, coherent bremsstrahlung). A light projectile colliding with a heavy target particle will be deflected (→ Multiple Scattering), but will lose little energy unless the collision is inelastic (→ Bremsstrahlung, Pair Production). A heavy projectile colliding with a light target will lose energy without being appreciably deflected.

The average energy loss of a hadron is mainly due to strong interactions, which eventually even destroy the particle (→ Calorimeter). Nevertheless, electromagnetic energy loss of hadrons is important, because the mean free path for strong interactions (→ Collision Length) is large.

Except when the projectile is highly relativistic, *ionization* is the main electromagnetic contribution to the energy loss. The mean energy loss (the stopping power) due to ionization is given by the *Bethe-Bloch formula* (→ [Barnett96], for more discussion [Leo94]).

$$-\mathrm{d}E/\mathrm{d}x = D\,\frac{Z}{A}\,\rho\,z^2\,\Phi(\beta)\,(1+\nu)\,,$$

with

$$\Phi(\beta) = \frac{1}{\beta^2}\left(\log\left(\frac{2m_\mathrm{e}c^2\gamma^2\beta^2}{I(1+\gamma m_\mathrm{e}/M)}\right) - \beta^2 - \frac{\delta}{2} - \frac{C}{Z}\right),$$

where

E	=	projectile energy
M	=	projectile mass
β	=	projectile velocity (in units of c)
γ	=	$1/\sqrt{(1-\beta^2)}$
z	=	projectile charge (in units of elementary charge)
x	=	path length
D	=	$4\pi r_\mathrm{e}^2\, m_\mathrm{e}c^2 N_\mathrm{A} = 0.30707$ MeV cm^2/mole
r_e	=	$2.817\,938\;10^{-13}$ cm = classical electron radius
m_e	=	$0.511\,003$ MeV/c^2 = electron rest mass
N_A	=	$6.022\;10^{23}$/mole = Avogadro's number
Z	=	atomic number of the medium
A	=	atomic weight of the medium [g/mole]
ρ	=	mass density of the medium [g/cm^3]
I	=	average ionization potential
δ	=	density correction
C	=	shell correction
ν	=	higher order correction.

The ionization energy loss is to a good approximation proportional to the electron density in the medium (given by $\rho Z N_A/A$) and to the square of the projectile charge, and otherwise depends mainly on the projectile velocity. It decreases with $1/\beta^2$ for increasing velocity until reaching a minimum around $\beta\gamma = 3$ to 4 (*minimum ionization*), then starts to rise logarithmically (*relativistic rise*) levelling off finally at a constant value (the *Fermi plateau*). The numerical value of the minimum ionization (more precisely: of minimum energy loss) is $\mathrm{d}E/\rho\mathrm{d}x \approx 2$ MeV cm^2/g.

The first expressions for energy loss are due to Bethe and Livingston, later Rossi gave more refined descriptions including various correction terms (→ [Rossi65], [Livingston37]). Sternheimer has worked in detail on the density effect which is at the origin of the Fermi plateau [Sternheimer71]. For a complete review, → [Fano63].

The formulae and in particular the corrections include absorber-dependent terms defying simple description (e.g. ionization potential and shell correction). Hence energy loss is usually given in graphical or tabular form (e.g. [Barnett96]). Extensive tables for the energy loss of p, K, π and μ in many materials have been computed by Richard and Serre ([Serre67], [Richard71]), where the Bethe–Bloch formula is also discussed with respect to the units used.

Measurements of energy loss, when giving enough care to calibration problems, can be used to identify particles if a simultaneous measurement of momentum is available. For details, → Ionization Sampling and [Allison91]. An example for measurements is the following diagram, obtained in a time projection chamber, taken from [Abreu96]:

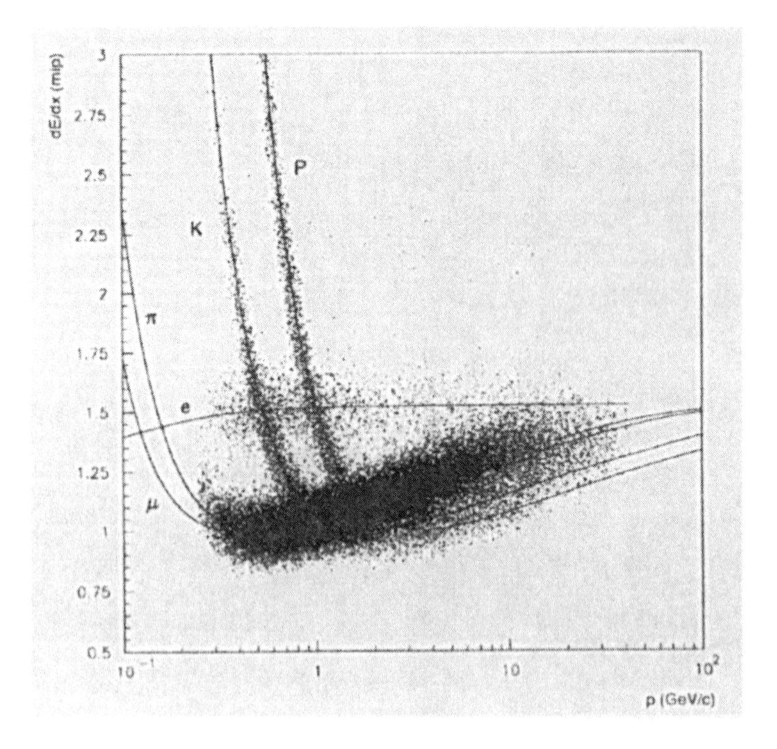

The ionization energy loss is statistically distributed around its mean value. The distribution, often referred to as *energy straggling*, is approximately Gaussian for thick absorbers, but develops asymmetry and a tail towards high energies for decreasing thickness; it becomes a Landau distribution (→) for very thin absorbers (→ [Leo94] or [Matthews81]).

Energy Resolution in Calorimeters. The ultimate limit for the energy resolution of a calorimeter is determined by fluctuations inherent in the development of showers, and by instrumental and calibration limits. The basic phenomena in showers are statistical processes, hence the intrinsic limiting accuracy, expressed as a fraction of total energy, improves with increasing energy as:

$$(\sigma/E)_{\mathrm{fluct}} \propto 1 / \sqrt{E} \, .$$

Over much of the useful range of calorimeters, this term dominates energy resolution.

There are other contributions than statistics, though: a second component is due to instrumental effects, being rather energy-independent (noise, pedestal); its relative contribution decreases with E:

$$(\sigma/E)_{\mathrm{instr}} \propto 1 / E \, .$$

This component may limit the low-energy performance of calorimeters.

A third component is due to calibration errors, non-uniformities and non-linearities in photomultipliers, proportional counters, ADC's, etc. This contribution is energy-independent:

$$(\sigma/E)_{\mathrm{syst}} = \mathrm{const.}$$

This component sets the limit for the performance at very high energies.

The two types of showers have markedly different characteristics:

a) *Electromagnetic Showers:* For electromagnetic showers (→) the intrinsic limitation in resolution results from variations in the net track length of charged particles in the cascade; for homogeneous shower counters

$$(\sigma/E)_{\mathrm{fluct}} \approx 0.005/\sqrt{E} \; [\mathrm{GeV}]$$

In sampling calorimeters, one has to add the sampling fluctuations:

$$(\sigma/E)_{\mathrm{samp}} \approx 0.04\ \sqrt{(1000\Delta E/E)}\ ,$$

with ΔE the energy loss of a single charged particle in one sampling layer. There are also fluctuations arising from the Landau distribution (→); a comparison can be found in [Fabjan91].
In practice, total energy resolution below the percent level for $E \leq 50$ GeV can be achieved routinely in homogeneous calorimeters; the same seems more like a very tough lower limit for sampling calorimeters. At low energies and for crystal calorimeters, total energy resolutions

$$(\sigma/E) \approx 0.025\ /\ {}^{4}\sqrt{E}\ [\mathrm{GeV}]$$

have been reported. For more quantitative values, → [Fabjan95a], [Gratta94].

b) *Hadronic Showers:* For hadronic showers (→) the intrinsic limitation is due to fluctuations in the fractional energy loss accounted for by the many interaction mechanisms leaving behind non-hadronic debris (including muons and the γ's and e^+/e^- from π^0 decays) and slow neutrons, along with fast hadrons. The fluctuations in these production processes, much larger than for electromagnetic processes, are the major ingredient of the final performance of a hadron calorimeter.
Intrinsic shower fluctuations are given by:

$$(\sigma/E)_{\mathrm{fluct}} \approx 0.45/\sqrt{E}\ [\mathrm{GeV}]$$

for uncompensated calorimeters, and with compensation for nuclear effects (→ Hadronic Shower, Compensating Calorimeter)

$$(\sigma/E)_{\mathrm{fluct}} \approx 0.25/\sqrt{E}\ [\mathrm{GeV}]\ .$$

Compared with the intrinsic fluctuations, sampling fluctuations are normally small:

$$(\sigma/E)_{\mathrm{samp}} \approx 0.09\sqrt{(1000\Delta E/E)}\ ,$$

with ΔE again the energy lost by a single charged particle in one sampling layer (note that $\Delta E/E$ is a very small number).
Note that these numbers refer to single hadronic particles; the σ/E for jets is typically higher by a factor 1.3 or more.
Hadronic showers can spread over a large volume; a major source of systematic errors, therefore, is the geometric limitation of a calorimeter. The resolution figures determined by intrinsic shower and sampling fluctuations will not be reached if showers are not

fully contained within the calorimeter volume. In practice some average fraction of the shower energy escapes through the sides (lateral leakage) or back (longitudinal leakage). While the corrections for longitudinal leakage are understood, and can partly be accounted for, corrections for lateral leakage need a careful inspection of the shower development and an estimate of the particle impact point.

More reading can be found e.g. in [Gordon95], [Fabjan91], [Wigmans91a], [Brau90].

Equations of Motion. The Lorentz force (→) causes a particle to bend in a magnetic and/or electric field. In most cases, the electric field is negligible; then the magnetic field is time independent, $\partial \boldsymbol{B}/\partial t = 0$. In this case, the energy E and momentum $p = |\boldsymbol{p}|$ are constants of motion. Let s be the path length, $\boldsymbol{r}$ the position of the particle, t the time, q the particle charge, c the speed of light; then the equation of motion can be written in various different ways, e.g.,

$$\mathrm{d}^2\boldsymbol{r}/\mathrm{d}t^2 = (qc^2/E)(\mathrm{d}\boldsymbol{r}/\mathrm{d}t) \times \boldsymbol{B} \ ,$$

the form which is most often used,

$$\mathrm{d}^2\boldsymbol{r}/\mathrm{d}s^2 = (q/p)(\mathrm{d}\boldsymbol{r}/\mathrm{d}s) \times \boldsymbol{B} \ ,$$

or in terms of $y' = \mathrm{d}y/\mathrm{d}x$, $y'' = \mathrm{d}^2y/\mathrm{d}x^2$, etc.,

$$\begin{aligned} y'' &= (q/p)\ s'(z'B_x + y'z'B_y - (1 + (y')^2)B_z) \\ z'' &= (q/p)\ s'\ (-y'B_x + (1 + (z')^2)B_y - y'z'B_z) \ , \end{aligned}$$

where $\mathrm{d}s^2 = |\mathrm{d}\boldsymbol{r}|^2 = \mathrm{d}x^2 + \mathrm{d}y^2 + \mathrm{d}z^2$ and

$$s' = \mathrm{d}s/\mathrm{d}x = (1 + (y')^2 + (z')^2)^{1/2} \ .$$

For singly charged particles, $|q|$ is the elementary charge e, which can be expressed in the usual *hybrid units* (→ Units) m or cm, s, GeV/c, T = tesla or G = gauss, as

$$e = 0.2998\ (\mathrm{GeV}/c)\,\mathrm{T}^{-1}\,\mathrm{m}^{-1} = 0.2998\ 10^{-6}\ (\mathrm{GeV}/c)\ \mathrm{G}^{-1}\,cm^{-1} \ .$$

These are three versions of the *equation of motion*, two of which do not involve time. The three versions are equivalent, even though the number of equations is apparently different; in each version, there are two independent second-order differential equations. For example, the three equations of the first version satisfy the identity

$$(\mathrm{d}^2\boldsymbol{r}/\mathrm{d}t^2)\cdot(\mathrm{d}\boldsymbol{r}/\mathrm{d}t) = 0 \; .$$

The second version has a simple geometrical interpretation. In fact, $\mathrm{d}\boldsymbol{r}/\mathrm{d}s$ is the unit vector tangent to the track, while $\mathrm{d}^2\boldsymbol{r}/\mathrm{d}s^2$ is a normal vector of length $1/\rho$, where ρ is the radius of curvature. Hence, if θ is the angle between $\mathrm{d}\boldsymbol{r}/\mathrm{d}s$ and $\boldsymbol{B}$, then

$$\begin{aligned} 1/\rho &= qB\ \sin\theta/p \; , \\ p\ \sin\theta &= qB\rho\ \sin^2\theta \; . \end{aligned}$$

$p\sin\theta$ is the momentum component normal to $\boldsymbol{B}$, while $\rho\sin^2\theta$ is the radius of curvature of the projection of the track onto a plane normal to $\boldsymbol{B}$.

The same equation can be given a different interpretation. When the particle travels an infinitesimal distance $\mathrm{d}s$, its track is deflected by an angle

$$\mathrm{d}\varphi = |\mathrm{d}^2\boldsymbol{r}/\mathrm{d}s^2|\,\mathrm{d}s = (q/p)\ B\sin\theta\,\mathrm{d}s \; .$$

where $B\sin\theta$ is the component of $\boldsymbol{B}$ normal to the track. Integration of this formula gives

$$\Delta\varphi = (q/p)\int B\sin\theta\,\mathrm{d}s \; .$$

Note that $\Delta\varphi$ so defined is usually only approximately equal to the total deflection angle, because rotation angles are only additive if the rotation axis is fixed.

The effect of a homogeneous field inside a rectangular box, say

$$\begin{aligned} B_x &= B_y = 0, \qquad B_z = B \quad \text{for} \quad x_1 < x < x_2 \; , \\ B_x &= B_y = B_z = 0 \quad \text{for} \quad x \le x_1, \quad \text{or} \quad x \ge x_2 \; , \end{aligned}$$

is to give a particle passing through at any angle a transverse *momentum kick*

$$\Delta p_T = \Delta p_y = (p_y)_2 - (p_y)_1 = -qB(x_2 - x_1) \; .$$

The homogeneous field in a box is a good model for many spectrometer magnets, if one replaces $B(x_2 - x_1)$ by the field integral $\int B\,\mathrm{d}s$ along a straight line perpendicular to the box containing the field. For example, if $\int B\,\mathrm{d}s = 2$ Tm (tesla × metre), the transverse momentum kick of a proton is $\Delta p_T = 0.6$ GeV/c. The variation of the integral $\int B\,\mathrm{d}s$ determines the precision of the model. With a formula containing only one parameter, one may get a precision of 10–20% in the momentum determination.

For detectors inside a homogeneous field B, one often uses the approximate relation between measured sagitta s, track length l, radius of curvature ρ and momentum p_{proj} (all quantities projected into the plane perpendicular to B):

$$s = \frac{l^2}{8\rho} = \frac{qBl^2}{8p_{\mathrm{proj}}} .$$

The precise solution of the Lorentz equations of motion (→ Trajectory of a Charged Particle) is found using methods of numerical integration (→ [Bock98]).

Errors in Track Reconstruction. The trajectory of a particle in a magnetic field is determined by five initial values, e.g.

$$x_0,\ y_0,\ (\mathrm{d}x/\mathrm{d}s)_0,\ (\mathrm{d}y/\mathrm{d}s)_0,\ (1/p)$$

(→ Trajectory of a Charged Particle). Therefore at least five measurements are necessary to reconstruct a track. In the absence of a magnetic field, four measurements are sufficient; the momentum remains unknown. Many methods are applicable to get an estimate of the track parameters; the most common method is the least squares method. In the case where the impact points of a particle on the detector surfaces can, in the neighbourhood of a given track, be approximated by a linear function of the initial values, the estimator found by the least squares method is the best possible linear estimator. If the measurement errors are Gaussian, the least squares estimator is also efficient (viz. has minimum variance). For most cases, the errors in track reconstruction as given by the theory of least squares estimation are quite representative for the achievable precision.

For simplification, let us consider two types of spectrometers:

a) a central spectrometer magnet with several detector arms: this setup is typical for high-energy fixed target experiments;
b) a set of equidistant detectors, all inside a magnetic field: this is typical for detectors in colliding beam experiments.

Ad a): Assuming a constant bending power $\int B\,\mathrm{d}l$ of the magnet, the *transverse momentum kick* given to a particle in the magnet can be approximated by (→ Lorentz Force)

$$p_{\mathrm{T}} = e\int B\,\mathrm{d}L \approx eBL ,$$

where p_T is the transverse momentum [GeV/c], e is the elementary charge [$e = 0.2998$ GeV/c $\mathrm{T}^{-1}\,\mathrm{m}^{-1}$], B the magnetic field [T], and L the length along the track [m]. For a particle with charge +1 and a magnet with $\int B\,\mathrm{d}l = 1$ Tm, $p_T = 0.2998$ GeV/c and the deflection angle α is given by

$$\alpha = p_\mathrm{T}/p = 0.2998/p\ .$$

With m available detectors and a symmetric spectrometer of length L, the theoretically best angular resolution is obtained by placing $m/4$ detectors at each end of the arm and $m/2$ detectors at the centre $L/2$. The obtainable resolution is then

$$|\Delta p/p| = |\Delta\alpha/\alpha| = (|p|/0.2998)(2\sigma/L)(4/\sqrt{m})\ ,$$

where σ is the error in an individual detector. This configuration of detectors, whilst optimizing precision, is a particularly unsuitable arrangement for finding the correct association of measurements, and is therefore not used in experiments with non-trivial track recognition problems.

Ad b): In central spectrometers all detectors are usually assumed to be inside a homogeneous magnetic field. This case is extensively discussed in [Gluckstern63]. The error of the reconstructed momentum in any projection is inversely proportional to the field in this projection and to the square of the projected track length L_p. Assuming the measured points to be equidistant and $m \gg 3$,

$$\Delta p/p = (\sigma/L_\mathrm{p}^2)(p_\mathrm{p}/0.2998B)\sqrt{[720/(m+6)]}\ .$$

m is the number of measurements and σ the error of a single measurement in this projection. For a fixed m and a given precision $\Delta p/p$ the spectrometer must grow in size with $\sqrt{p}$. If m grows linearly with L_p, $\Delta p/p$ is asymptotically proportional to $p_\mathrm{p} m^{-5/2}$. Note that $p_\mathrm{p}/L_\mathrm{p}^2 = p/(L^2\cos\lambda)$, where L is the track length in space and λ the projection angle.

If half of the measurements are assumed to be in the centre of the track, and one fourth each at the ends, the momentum error is substantially improved to

$$\Delta p/p = (\sigma/L_\mathrm{p}^2)(p_\mathrm{p}/0.2998B)\sqrt{[256/(m+2)]}\ .$$

This is, again, a hypothetical arrangement of detectors, as it is unsuitable for recognizing tracks and also difficult to install. Note that even this formula gives a precision worse than the lever arm spectrometer by a factor of 2.

At high energies, the momenta measured in magnetic fields by position detectors will have large errors, and calorimetric measurements are preferred, particularly for electrons.

At low energy, a precision limit is set by multiple scattering (→) and the optimization becomes definitely more complicated, as it will depend on the distribution of the scattering material (continuous or discrete) and on the momentum spectrum of the particles. Whereas the relative momentum error from position measurement errors is given by the proportionality

$$(\Delta p/p)_{\text{meas}} \approx p_{\text{p}}/L_{\text{p}}^2 \, ,$$

the corresponding term from multiple scattering comes out to be

$$(\Delta p/p)_{\text{multsc}} \approx \sqrt{(p^2+m^2)}/(\beta p\sqrt{L_{\text{p}}}) \, ,$$

where a uniform spacing of a constant number of measured points is assumed. For the full mathematical treatment of multiple scattering, → [Gluckstern63] and [Pentia96] and references therein.

If the direction of a straight track (no magnetic field) is calculated from

measurements c_1 for transverse coordinates y_1 in detectors positioned at longitudinal coordinates x_1, the least squares fit to the equation $y_1 = ax_1 + b$ gives

$$a = \left(n\sum x_1c_1 - \sum x_1 \sum c_1\right)/D$$

$$b = \left(\sum c_1 \sum x_1^2 - \sum x_1 \sum x_1c_1\right)/D$$

$$D = n\sum x_1^2 - \left(\sum x_1\right)^2 \, .$$

The errors follow from error propagation, using the relation between a, b and the measurements c_i.

More details, and consideration of more complicated setups, including vertex chambers, can be found in [Blum93].

Exclusive Measurement of Interactions. A measurement of particle interactions in which all participating particles are identified and measured or computed in momentum (for the opposite, → Inclusive Measurement of Interactions). Exclusive measurements including all particles of an interaction are usually possible only at relatively low laboratory energies and for simple interaction types like two- and three-body final states.

Fermi Plateau. A range of high energies, $\beta\gamma = p/mc \geq 100$, where the energy loss (→) of a particle traversing a medium no longer increases with increasing particle energy. Up to this plateau, the loss of energy (→ Energy Loss) of a traversing particle increases logarithmically with energy (*relativistic rise*). The onset and the height of the Fermi plateau are due to the density correction cancelling the relativistic rise; they vary for different media. Calculations and measurements are published for a wide range of solids, liquids and gases, and show good agreement; → e.g. [Cobb76], for gases [Burq81], [Walenta79a] or [Allison76].

Feynman Diagram. Feynman diagrams are pictorial representations of interactions between quantized fields. In a Feynman diagram quanta are represented by lines (edges) which interact at vertices (nodes). Usually, the time coordinate is represented left to right, the space coordinate up-down. For an example, → Deep Inelastic Scattering Variables.

Feynman diagrams are translated into scattering amplitudes by assigning couplings and propagator terms to nodes and edges (→ [Hooft73]).

Feynman x. The variable $x = P_L/P_{L(\max)} = 2P_L/\sqrt{s}$ was introduced by Feynman [Feynman69] as scaling variable (→) in the discussion of inclusive hadronic interactions at large energies. P_L is the longitudinal momentum of a particle, $P_{L(\max)} = \sqrt{s}/2$ is the maximum allowed P_L, and $\sqrt{s}$, the total centre-of-mass energy of the interaction. Also used sometimes is the *transverse* x defined by $x_T = 2P_T/\sqrt{s}$ (with P_T the transverse momentum).

In hadron-hadron collisions, x_1 and x_2 are often used to designate the energy fraction carried by the two colliding partons. The parton–parton system is defined (assuming massless partons and head-on collisions) by a centre-of-mass energy $\sqrt{(sx_1x_2)}$. The probability density function of the x_i is called the *structure function*.

Field Shaping. The usage of surfaces (wires, planes, strips) with controllable potential configuration to get a desired shape for the electric field in a given volume, typically in drift chambers (→). This includes, of course, the introduction of additional wires or grids, in order to optimize the field, e.g. to get a uniform drift field in the cells of a drift chamber, or a simple relation between drift time and distance, independent of track parameters. *Field shaping wires* are in principle all those wires in multiwire chambers (→) which contribute to the shape of the electric field in which the electrons drift towards a sense wire. This includes, of course, the cathode wires. Sometimes, wires with different potential are used to obtain optimal conditions; resistive voltage dividers allow one to obtain the desired potential. In drift chambers one also talks about the cathode wires as field shaping wires and about the (thicker) wires that alternate with the sense wires as field wires. When executed as conductive strips on printed circuit boards, usually in narrow bends and long straight sections, the name *racetracks* is also in use. Separate grids are introduced in time projection chambers (→) and time expansion chambers (→), to gate out ions and control the drift region.

For a fixed geometry, the solutions to the differential equations for electrostatic fields can be found in textbooks of classical electrodynamics (for a few basic formulae, → [Barnett96]). For literature on detectors, → [Blum93], [Sauli91], [Sill90a], [Sill90b], and [Beingessner80].

Field Wire. → Field Shaping

Four-momenta. Energy and vector momenta of a particle or set of particles can be combined into a four(dimensional)-vector p. The components are

$$\begin{aligned} p^0 &= p_0 = E \quad \text{(Energy)} \\ p^1 &= -p_1 = p_x \\ p^2 &= -p_2 = p_y \\ p^3 &= -p_3 = p_z \end{aligned}$$

where p_x, p_y and p_z are the momentum components along a system of orthogonal axes. Then if the particle has mass m,

$$m^2 = E^2 - |\mathbf{p}|^2 = (p^0)^2 - (p^1)^2 - (p^2)^2 - (p^3)^2 \; .$$

Using the convention of summing over repeated indices, this last line can be written as

$$m^2 = g_{\mu\nu} p^\mu p^\nu$$

where $g_{\mu\nu}$ is a tensor, the metric tensor of the Minkovski or Lorentz geometry

$$\begin{pmatrix} +1 & . & . & . \\ . & -1 & . & . \\ . & . & -1 & . \\ . & . & . & -1 \end{pmatrix}$$

in which dots indicate zeros. p^μ with an upper index is called a contravariant four-vector, while p_μ with a lower index is covariant; the relation is that $p_\mu = g_{\mu\nu}\, p^\nu$.

The principles of conservation of linear momentum and of energy together give conservation of four-momentum, in any elementary particle interaction or decay. For any system of particles an effective mass can be defined

$$M^2 = g_{\mu\nu} p^\mu p^\nu = E^2 - |\boldsymbol{p}\,|^2$$

where $p = (E, \boldsymbol{p})$ is the total four-momentum. The effective mass is a relativistic invariant and is conserved in any interaction. Thus if a particle of mass M decays into several other particles, their effective mass will be M.

The scalar product of any pair of four-vectors is defined by

$$\begin{aligned} p \cdot q &= g_{\mu\nu} p^\mu q^\nu = p^\mu p_\mu \\ &= p^0 q^0 - p^1 q^1 - p^2 q^2 - p^3 q^3 = p^0 p_0 + p^1 p_1 + p^2 p_2 + p^3 p_3 \,. \end{aligned}$$

Fragmentation Function. The probability density function of a characteristic variable describing the hadronization of jets, e.g. longitudinal momenta of hadrons inside a quark or gluon jet. Typically, the variables used are $x_{\mathrm{P}} = p_{\mathrm{had}}/p_{\mathrm{jet}}$ or $z = p_{\mathrm{L}}/p_{\mathrm{jet}}$ with p_{L} being the hadron momentum along the jet axis. The extreme values for both variables are zero and one. The fragmentation functions measured in e^+e^- or $\bar{p}p$ interactions are characterized by a peak at zero and a fast experimental falloff towards higher values.

The measured fragmentation functions show clear differences between quark and gluon jets: gluon jets have higher particle multiplicities, but their energy and energy fraction is lower: the fragmentation is *softer*. → e.g. [Buskulic96], [Gary94].

Fragmentation Region. The small-angle (centre of mass) region of an interaction. Particles in the fragmentation region have momenta similar to the incident or target particle. Consequently one speaks about the *beam fragmentation* region, sometimes defined over a given range of rapidity (→) like $y_{\mathrm{max}} - \Delta < y \leq y_{\mathrm{max}}$ where $y_{\mathrm{max}} =$

$\log(\sqrt{s/m})$ and $\Delta \approx 2$, or about the *target fragmentation* region (defined similarly for negative y).

Gaseous Detectors, Operational Modes. The charge collected by the anode of a chamber depends on the intensity of the electric field applied to the chamber. At some low voltage, the recombination of electrons and ions is overcome, but no gas multiplication occurs; a detector in this mode is insensitive to the voltage, and is called an *ionization chamber*; the output signal is weak, and corresponds to the number of primary electrons.

As the voltage is increased, which happens in the fields of wire chambers, the primary ionization electrons cause *electron avalanches* to form: the accelerating electric field is high enough to impart to the electrons, generated by the primary ionization in the gas, an energy higher than the first ionization potential of the gas. These electrons then produce ion–electron pairs while continuing along their path; the secondary electrons may, in turn, form further pairs, and the phenomenon is called *gas multiplication*. Eventually, the freed electrons drift towards the anode and produce an analogue signal that can be used for position and energy loss measurement. Most wire chambers work in this *proportional mode*, viz. the signals recorded by the detector are much higher and still proportional to the energy loss $\mathrm{d}E/\mathrm{d}x$ of the traversing particle. In most practical chambers, the electric field close to the thin (20–30 µm) anode wire has a high gradient, so that a multiplication factor of 10^5 to 10^6 is reached, with multiplication occuring mostly very close to the wire, where the field is strongest.

Strict proportionality assumes that space charge (due to the longer-lived positive ions) and induced effects remain negligible, compared to the external field. At higher electric fields, or in a high flux of charged particles, the space charge effects alter the effective electric field, the chamber works in the mode of *limited proportionality*: the signal is no longer strictly proportional to the energy loss of the particle; the relation between collected charge and $\mathrm{d}E/\mathrm{d}x$ can still be put to use, though.

Further increase of the electric field eventually leads to electric breakdown of the gas. This takes place when the space charge inside the avalanche is strong enough to shield the external field. A recombination of ions then occurs, resulting in photon emission and in secondary ionization with new avalanches beyond the initial one. If the process propagates (backwards, from the avalanche tail) until an ion column links anode and cathode, a spark discharge will eventually occur, and a chamber or counter is said to operate in the *Geiger–Müller mode.*

In the *limited Geiger mode*, this discharge is not allowed to happen, which can be achieved by adding quenching agents to the gas (→ Gas Mixtures in Gaseous Detectors); output pulses at the anode are much higher in this mode than in the proportional mode. The process of spark discharge can also be stopped by manipulating the electric field: if only short (a few ns) pulses of high voltage are applied, short discharges develop from the ion trail of a crossing particle (*streamers*), and a track image can be obtained by photography (*streamer chamber*).

A similar effect as for the limited Geiger mode can be obtained using thick (50–100 μm, as opposed to the usual 20–30 μm) anode wires [Brehin75] without using quenchers. This mode of operation, attractive because of its high mechanical reliability due to the thick wires, is called the *limited streamer mode.*

For more details, → [Sauli91], [Blum93].

Gas Mixtures in Gaseous Detectors. Avalanche multiplication is essential in all gaseous detectors, in order to produce an electrical signal of sufficient amplitude. In principle, all gases can be used for generating electron avalanches, if the electric field near the (sense) wire is strong enough. However, depending on the mode of operation (→ Gaseous Detectors, Operational Modes) and the intended use of the chambers, specific requirements towards, e.g. signal proportionality, high gain, good drift properties, or short recovery times, limit the choice of gases or gas mixtures.

Multiplication occurs in noble gases at lower fields than in gases with complex molecules; the addition of other components increases the threshold voltage. This suggests a noble gas as the main component of a chamber gas. Noble gases do not, however, allow operation at high enough gas gain without entering into a permanent discharge operation; the atoms excited during the avalanche process return to the ground state emitting photons at high enough energies to initiate a new avalanche in the gas or around the cathode. The latter may also be induced by the neutralization of ions that travel to the cathode. This problem is solved by the addition of a *quenching gas* which absorbs energetic photons; usually this is an organic gas like isobutane $(CH_3)_2CHCH_3$. Most organic compounds in the hydrocarbon and alcohol families are efficient in absorbing photons in the relevant energy ranges. The molecules dissipate the excess energy either by elastic collisions, or by dissociation into simpler radicals. Even a small amount of a polyatomic quencher added to a noble gas changes completely the operational characteristics of a chamber, and may allow gains in excess of 10^6 to be obtained before discharge.

Classical gas mixtures for proportional counters are P10 (90% Ar+10% CH_4) and for (proportional) multiwire chambers (→), MWPCs for short, the "magic gas" mixture: 75% Ar + 24.5% isobutane + 0.5% freon.

Different requirements apply to chambers with long drift time; they include (besides the properties of gases of MWPCs) particularly good drift properties: gas purity is important, and special attention must be given to the drift velocity. If the chamber is to operate at high counting rates, the drift velocity should be high, to avoid losses due to dead time. For better spatial resolution, drift velocities should be lower, to minimize the influence of timing errors on position resolution. Characteristic for this category are gases like dimethylether (DME) or CO_2.

In microstrip gas chambers (→), MSGCs for short, the gas mixtures should have the following characteristics (→ [Schmitz94]):

- high primary ionization density of the gas mixture, to reach full efficiency in a thin layer of gas;
- high electron drift velocity, to achieve a large signal and keep the detector occupancy low in a high flux environment;
- high maximum gas amplification factor, to match the noise level of the electronics;
- the gas mixture should not cause fast detector aging;
- for use in magnetic fields, the gas should have a low Lorentz angle (→), which is achieved by a high electric field in the drift region.

Typical mixtures proposed for MSGCs are of the type Xe + CO_2 + DME. Detailed studies can be found in [Geijsberts92], [Beckers94].

For introductory reading, → [Blum93]. More about gases in wire chambers can be found in [Sauli91] or [Peisert84]. [Va'vra92] discusses in detail the simulation of the behaviour of gas mixtures on computers. On aging of wire chambers, → Radiation Damage in Gaseous Detectors.

Gas Multiplication. → Gaseous Detectors, Operational Modes

Geiger Mode. → Gaseous Detectors, Operational Modes

Hadron Calorimeter. Calorimeter (→) optimized for incident hadrons, usually placed behind an electromagnetic calorimeter which fully contains electromagnetic showers (→ Hadronic Shower, Electromagnetic Shower).

Hadronic Shower. The hadronic showering process is dominated by a succession of inelastic hadronic interactions. At high energy, these are characterized by multiparticle production and particle emission originating from nuclear decay of excited nuclei. Due to the relatively frequent generation of π^0's, there is also an electromagnetic component present in hadronic showers.

Secondaries are mostly pions and nucleons. The hadronic multiplication process is measured at the scale of nuclear interaction length (→) λ_1, which is essentially energy-independent.

Intrinsic limits on the energy resolution of hadronic calorimeters are:

- A fluctuating π^0 component among the secondaries which interacts electromagnetically without any further nuclear interaction ($\pi^0 \rightarrow \gamma\gamma$). The average fraction of π^0's is given by $\pi_0/\text{all} \approx 0.10 \log(E)$ [E in GeV]. Showers may develop with a dominant electromagnetic component.
- A sizeable amount of the available energy is converted into excitation and breakup of nuclei. Only a small fraction of this energy will eventually appear as a detectable signal and with large event-to-event fluctuations.
- A considerable fraction of the energy of the incident particle is spent on reactions which do not result in an observable signal. Such processes may be energy leakage of various forms, like:

  backscattering and other albedo processes,
  leakage due to μ, ν or slow neutrons,
  nuclear excitation, nuclear breakup, nuclear evaporation.

The average ratio between signals from electromagnetic and hadronic particles of the same incident energy is calorimeter- and energy-dependent; for a non-compensating calorimeter, one has typically

$$e/h \approx 1.1\text{–}1.35\,.$$

For possible improvements, → Compensating Calorimeter.

At high energies some characteristic numbers of hadronic showers can be described by a simple parameterization in terms of λ ($\rightarrow$ [Fabjan82]:

shower maximum:

$$l_{\max} \approx [0.6 \ \log(E) - 0.2]\lambda \ , \quad \text{with } E \text{ in GeV} \ ;$$

shower depth for 95% longitudinal containment:

$$l_{95\%} \approx l_{\max} + 4E^{a}\lambda \ , \quad \text{with } E \text{ in GeV and } a = 0.15 \ ;$$

shower radius for 95% radial containment:

$$R_{95\%} \approx \lambda \ .$$

Lengths are measured from the calorimeter face.

For the average differential energy deposit over the volume of the cascade an acceptable longitudinal parametric approximation is given in [Bock81]:

$$\mathrm{d}E = k[wt^{a-1}\,\mathrm{e}^{-bt} + (1-w)l^{c-1}\,\mathrm{e}^{-dl}]\,\mathrm{d}s$$

with

$$\begin{aligned} t &= \text{depth starting from shower origin [in } X_0] \\ l &= \text{depth starting from shower origin [in } \lambda_l] \\ \mathrm{d}s &= \text{step in depth} \\ a, b, c, d &= \text{parameters fitted from data} \\ w &= \text{relative weight of the electromagnetic component} \\ 1-w &= \text{relative weight of the hadronic component} \\ & \quad\ \text{of the shower} \\ k &= \text{normalization.} \end{aligned}$$

A simple formula for the average lateral hadronic shower development is not in common use; double exponentials or combinations of exponentials with Gaussian curves have been applied with success ($\rightarrow$ [Acosto92]); superposition of two Breit–Wigner distributions has also been experimented with ([Durston93]). Average energy depositions allow one, in principle, to give an (average) energy density in three dimensions, which can be mapped onto any calorimeter cell structure; fluctuations can be added, but are a poor approximation to the reality of showers. A more realistic parameterization of shower fluctuations has been proposed by [Iaselli92]; in that model, the mean μ

and standard deviation σ, and their correlations, are parameterized for the energy in each layer of a sampling calorimeter. While this results in realistic showers, the method is applicable only for a specific calorimeter, and requires substantial preprocessing of fully simulated (or measured) showers.

Contrary to electromagnetic showers, which develop in subnanosecond time, the physics of hadronic showers is characterized by different time scales, the slowest of which (de-excitation of heavy nuclei) may reach a microsecond (→ [Caldwell93]).

Hadronization. → Jet Variables

Half Life. The time $\tau_{1/2}$ in which a radiating body decreases in intensity by a factor 2, or the time in which half of a sample of decaying particles will indeed decay. The half life is related to the mean life τ by

$$\tau_{1/2} = \tau \log 2 \approx 0.69315\,\tau.$$

→ also Attenuation.

Hermeticity. A term used in calorimeters to describe maximal coverage for particles under all emission angles, combined with minimal leakage. A hermetic calorimeter allows the measurement of energy in all directions so as to infer, by forming vector sums, the energy that escapes unseen in the form of neutrinos. → Calorimeter.

Hodoscope. A combination of multiple detector elements arranged in space and connected by logic circuitry such that particle tracks can be identified (the literal translation from the Greek is "pathviewer"). Most often, hodoscopes are used for triggering purposes; they are based on fast detectors, usually scintillation counters (→) with very short output pulses.

Inclusive Measurement of Interactions. An inclusive measurement of a particle interaction is a partial measurement. Only a few produced particles, sometimes only one, are singled out for identification and measurement, ignoring the details of all other interaction products. Reactions can thus be written

$$A + B \rightarrow S_{\text{measured}} + \text{anything}\ .$$

Inclusive measurements dominate at high energies, where the separation of tracks and particle identification become difficult, even when using the most advanced detectors. Triggers in complex interactions are necessarily inclusive: the signature of interesting physics will be defined in terms of few phenomena, like high-p_T leptons or jets, disregarding the rest of the interaction.

Interaction Length. The mean free path (→) of a particle before undergoing an interaction that is neither elastic nor quasi-elastic (diffractive), in a given medium, usually designated by λ. The relevant cross-section is $\sigma_{\text{tot}} - \sigma_{\text{el}} - \sigma_{\text{diff}}$. For some numerical values, → Collision Length or [Barnett96].

Invariant Cross-Section. The cross-section $E\,\mathrm{d}\sigma/\mathrm{d}\boldsymbol{p}$ is called invariant because $\mathrm{d}\boldsymbol{p}/E$ remains invariant under Lorentz transformations. Inclusive cross-sections are commonly given as invariant cross-sections, e.g. in terms of rapidity y and transverse momentum p_T. Integrated over the azimuth, the invariant cross-section in these variables is $(1/\pi)\,\mathrm{d}\sigma/(\mathrm{d}y\mathrm{d}p_T^2)$. For other invariant cross-sections, → [Barnett96].

Ionization Chamber. A chamber operating at a voltage lower than needed for the onset of proportional operation (it collects the total primary ionization); → Gaseous Detectors, Operational Modes.

Ionization Sampling. The measurement of the energy loss $\mathrm{d}E/\mathrm{d}x$ of a charged particle, on many points along its trajectory; usually in order to determine its mass: an estimate of $\mathrm{d}E/\mathrm{d}x$ combined with momentum measurement, will allow one to put limits on the particle mass. Ionization sampling is often combined with the measurement of the particle position, but may also be left to devices dedicated to the measurement of $\mathrm{d}E/\mathrm{d}x$.

Typically, ionization sampling is done over small amounts of lost energy as in the gas of a drift chamber or in a bubble chamber liquid. Due to the fluctuations in the energy loss in thin slices, it is important to obtain a large number of samplings. It is then the statistical distribution of $\mathrm{d}E/\mathrm{d}x$ values measured for the same track which allows an estimation of the velocity β, on which energy loss (→) depends. In analysing the sample of local $\mathrm{d}E/\mathrm{d}x$ values, care must be taken to use a sensible estimator: due to the tail in the Landau distribution, a simple mean value will be a bad estimator. Mostly, one uses either a truncated mean (like eliminating the 20% of highest individual measurements), or resorts to a full maximum likelihood treatment.

Ample discussion on optimizing detectors and readout for ionization sampling, and a vast amount of literature is also found in review papers, e.g. [Lehraus83], [Allison91]. [Blum93] discusses in detail the collection and analysis of ionization samples in drift chambers.

Jet Chamber. A drift chamber made of multiple cells of moderate size, named thus because of the optimal two-track resolution as needed in jets; → Drift Chamber.

Jet Variables. In many collisions, observable secondary particles are produced in highly collimated form, called particle *jets*. This is a consequence of the *hadronization* of partons (quarks or gluons) produced in hard collisions. Jets for a given initial parton can vary widely in shape, particle content, and energy spectrum; there is, of course, also substantial blurring due to instrumental effects: the finite resolution and granularity of detectors (calorimeter cells and muon measurements), and escaping neutrinos.

The earliest evidence for jets was in e^+e^- collisions (SLAC and DESY), producing secondary hadrons; subsequently, they were also observed in hadronic collisions (e.g. UA experiments and ISR at CERN). Frequently, two main jets are observed which dominate the energy balance of the collision; in hadronic collider events, the balance is observed only laterally, due to the difficulty of observing at large (absolute) rapidity, and due to the structure function (→), which leaves the hard quark encounter with a longitudinal boost. Often, the main jets are accompanied by one or more broader jet(s), interpreted as radiated gluons. The following scalar jet variables were used in the early jet studies, describing mostly a two- or three-jet situation from e^+e^- events.

a) Sphericity $\equiv$ $(3/2)\min\left(\sum \boldsymbol{p}_\mathrm{T}^2/\sum \boldsymbol{p}^2\right)$
   p_T is the transverse momentum perpendicular to a unit vector $\boldsymbol{n}$, the sums are over all particles of the reaction, and the minimum is formed with respect to $\boldsymbol{n}$.
b) Thrust $\equiv$ $2\max(\sum_1 \boldsymbol{p}_\mathrm{L}/\sum|\boldsymbol{p}|)$
   p_L is the longitudinal momentum along a unit vector $\boldsymbol{n}$. Summation is over all particles for $|p|$, over those with $\boldsymbol{p}\cdot\boldsymbol{n} > 0$ for p_L; the maximum is formed with respect to $\boldsymbol{n}$. For the hadronic system

in deep inelastic scattering, a *current thrust* has been proposed ($\rightarrow$ [Webber95]) which is the thrust evaluated in the *Breit frame*, defined to be the frame in which the momentum transfer q is spacelike and along the longitudinal ($z-$) axis ($\rightarrow$ Deep Inelastic Scattering Variables).

c) Spherocity $\equiv (4/\pi \min(\sum |p_{\mathrm{T}}|/\sum |p|)^2$
p_{T} is the transverse momentum perpendicular to a unit vector $\boldsymbol{n}$, the sums are over all particles, and the minimization is with respect to $\boldsymbol{n}$

d) Triplicity $\equiv \max(\boldsymbol{n}_1 \sum_1 \boldsymbol{p} + \boldsymbol{n}_2 \sum_2 \boldsymbol{p} + \boldsymbol{n}_3 \sum_3 \boldsymbol{p}) / \sum |\boldsymbol{p}|$
Here, a general classification into three classes must take place; in each class i, $\boldsymbol{n}_i$ describes the axis and particles are associated to the class for which $\boldsymbol{p} \cdot \boldsymbol{n}$ is largest. The maximum must be found over all possible $\boldsymbol{n}_1, \boldsymbol{n}_2, \boldsymbol{n}_3$.

e) Planarity $\equiv \max(\sum p_{\mathrm{T1}}^2 - \sum p_{\mathrm{T2}}^2)/\sum p_{\mathrm{T}}^2$
Here, p_{T1} and p_{T2} are defined to be axes of a Cartesian coordinate system whose third axis is p_{L}. The variable indicates how well a reaction satisfies the assumption of being in a plane. The maximum is found with respect to the plane orientation. Maximization gives the same result if $\sum p_{\mathrm{T1}}^2$ is maximized. The solution is therefore given by the principal axes (obtained in "principal component analysis"), and the planarity is the complement of a two-dimensional equivalent to sphericity (called *circularity*). If p_{L} is along x, then the direction for p_{T1} is obtained by rotating in the y-z plane through an angle α given by

$$\begin{aligned}
\tan(2\alpha) &= \left(2\sum(p_y p_z)\right) / \left(\sum p_y^2 - \sum p_z^2\right) \\
p_{\mathrm{T1}} &= p_y \cos\alpha + p_z \sin\alpha \\
p_{\mathrm{T2}} &= p_z \cos\alpha - p_y \sin\alpha \,.
\end{aligned}$$

As maxima and minima differ by 90° in α, solving the above equation for an α in the range, say, from 0 to $\pi/2$, will still require deciding whether a maximum or minimum has been found. Obviously, $p_{\mathrm{T1}}^2 + p_{\mathrm{T2}}^2 = p_y^2 + p_z^2 = p_{\mathrm{T}}^2$.

The variables a) to c) all describe two-jet situations, with the jets back to back ($p_{\mathrm{L}} = 0$). d) has been used for a general three-jet situation. A detailed discussion of these variables can be found in [Brandt79]. The planarity e) describes a four-jet situation, if pairs of jets are correlated, e.g. a hard-scattered and a spectator system. The scattered system may have p_{L} different from zero, but will be in a plane with the spectator jets and back to back in the transverse projection. A more recent discussion on e^+e^--produced jets can also be found in [Akrawy90].

Jets are not well-defined phenomena, hadrons in the final state not being rigorously associated to the partons (which in turn also undergo some final-state interactions before they hadronize). In hadron–hadron collisions, jets are particularly difficult to separate: there is an *underlying event* due to the remnants from the quarks not participating in the hard interaction, and at high luminosity, there may be *pileup* of multiple collisions that cannot be separated in time (e.g. at the planned Large Hadron Collider).

In jet analysis of more recent dates, therefore, simple jet variables have been somewhat abandoned; jets are analysed in *cones* defined by a cutoff (typically between 0.5 and 1.0) in an angular radius $R = \sqrt{\Delta\eta^2 + \Delta\varphi^2}$ around a jet axis defined in various ways. Jets thus found are often subsequently contracted, viz. combined into fewer jets, using some clustering algorithm.

The jet structure is studied as a function of the jet radius; particle multiplicities, rapidity distributions, fragmentation functions (→), and others (and again their gradients, when varying the cone radius) are eventually compared to the values obtained from phenomenological Monte Carlo programs; → e.g. [Ellis91], [Bethke91], [Buskulic96], [Varelas96].

Some specific variables are used for multijet events: in three-jet events, labelling the outgoing jets 3, 4, 5 after ordering them by decreasing energy (jet 3 is thus the highest-energy or *leading* jet), one uses

f) the Dalitz variables:

$$X_3 = 2E_3/M, \quad \text{and} \quad X_4 = 2E_4/M$$

with M the three-jet effective mass;

g) the scattering angle of the leading jet:

$$\cos\Theta_3 = \frac{\boldsymbol{p}_{\text{beam}} \cdot \boldsymbol{p}_3}{|\boldsymbol{p}_{\text{beam}}||\boldsymbol{p}_3|} ;$$

h) the angle in the three-jet rest frame between the plane containing the leading jet and the beam, and the three-jet plane:

$$\cos\Psi_3 = \frac{(\boldsymbol{p}_3 \times \boldsymbol{p}_{\text{beam}}) \cdot (\boldsymbol{p}_4 \times \boldsymbol{p}_5)}{|\boldsymbol{p}_3 \times \boldsymbol{p}_{\text{beam}}||\boldsymbol{p}_4 \times \boldsymbol{p}_5|} .$$

In these formulae, $\boldsymbol{p}_{\text{beam}} = \boldsymbol{p}_1 - \boldsymbol{p}_2$, refers to the effective parton beam, i.e. the difference between the momenta of the colliding partons. For higher jet multiplicities, one often uses variables reducing to three jets. For details and results, → e.g. [Abe92], [Acton93].

K-factor. This denomination is used in many contexts, the K standing for (the limits of) knowledge. Also called occasionally *fudge* factor, this is in general a way to express unknown (or difficult-to-express) effects by a correction factor. In high-energy physics, a K-factor is often used in comparing cross-sections calculated up to leading order (LO), to the same up to next-to-leading order (NLO) ($K = \sigma_{\text{NLO}}/\sigma_{\text{LO}}$), or in comparing observed values of cross-sections to those calculated ($K = \sigma_{\text{obs}}/\sigma_{\text{LO}}$).

Knock-on Electron. Also called *delta rays*, knock-on electrons are emitted from atoms by the passage of charged particles through matter. Any charged particle traversing a medium transfers energy to that medium via the process of ionization or excitation of the constituent atoms. Due to the statistical fluctuations in energy loss, there is some probability of transmitting energy in excess of a few keV; the precise cutoff energy has to be defined as a function of the detector: they become detectable tracks. Knock-on electrons have enough energy to produce, themselves, fresh ions in traversing the medium (secondary ionization).

The probability density function for the energy transfer is approximately given by $P(E)\,\mathrm{d}E = K\,\mathrm{d}X/E^2$, with X the particle path length, and E the kinetic energy; for a more detailed discussion of the energy distribution, → [Barnett96].

In cosmic ray physics using photographic emulsions, δ-rays served to determine the charge of the observed particle. In bubble chamber physics they were used for particle identification, which is possible because the kinematics for δ-ray production vary drastically with the mass of the traversing particle. Also, the δ-ray laboratory angle of emission differs for different particles. δ-rays up to a few keV are emitted more or less perpendicularly to the incident track, but their mean free path is only of the order of a few microns even at atmospheric pressure.

In wire chambers, knock-on electrons can distort the signal recorded on the sense wire, and cause occasional outliers (viz. very large charges), particularly in drift chambers.

KNO-Distribution. The multiplicity of particles created in high-energy collisions follows a distribution with a long tail, qualitatively similar to the distribution of energy loss (Landau distribution). The curve and particularly its scaling properties have been discussed by Koba, Nielsen and Olesen [Koba72], from whose initials its name has been derived. Parameterization attempts for multiplicity distributions have been discussed, and KNO scaling violations reported, e.g. [Friedländer91], [Alner87].

Landau Distribution. The fluctuations of energy loss by ionization of a charged particle in a thin layer of matter was first described theoretically by Landau [Landau44]. They give rise to a universal asymmetric probability density function characterized by a narrow peak with a long tail towards positive values (due to the small number of individual collisions, each with a small probability of transferring comparatively large amounts of energy. The mathematical definition of the probability density function is

$$\varphi(\lambda) = \frac{1}{2\mathrm{i}\pi} \int_{c-\mathrm{i}\infty}^{c+\mathrm{i}\infty} \mathrm{e}^{s \log(s) + \lambda s} \, \mathrm{d}s \ ,$$

where λ is a dimensionless number and is proportional to the energy loss, and c is any real positive number. Other expressions and formulae are indicated in [Kölbig83] (with program implementations) and [Leo94]; e.g.

$$\varphi(\lambda) = \frac{1}{\pi} \int_0^\infty \mathrm{e}^{-s \log(s) - \lambda s} \ \sin(\pi s) \, \mathrm{d}s$$

Programs also exist for the distribution (integral function) of $\varphi(\lambda)$, for its derivative, and for the first two moments.

[Moyal55] has given a closed analytic form

$$\Psi(\lambda) = \sqrt{\frac{\mathrm{e}^{-(\lambda + e^{-\lambda})}}{2\pi}}$$

with

$$\begin{aligned} \lambda &\doteq R\,(E - E_\mathrm{p}) \\ E_\mathrm{p} &= \text{most probable energy loss} \\ R &= \text{constant depending on the absorber} \ . \end{aligned}$$

This curve reproduces the gross asymmetric features of the Landau distribution and avoids the pitfalls of numerical integration; it is, however, too low in the tail and unrelated to $\varphi(\lambda)$ as defined above.

On the other hand, Moyal's curve is useful in some situations: as the layer over which energy loss is integrated becomes thicker, the tail of the Landau distribution (according to the *central limit theorem* of statistics) has a tendency to diminish:[1] the Landau distribution becomes the more general *Vavilov distribution.* Vavilov introduced the additional parameter $k = E_{\mathrm{av}}/E_{\mathrm{max}}$, with E_{av} the average energy loss over the layer, and E_{max} the maximum energy loss in a single collision. The Landau distribution is the limiting case for $k = 0$ and a good approximation for $k < 0.01$; the Gaussian is the limiting case for $k = \infty$ and a reasonable approximation for $k > 10$. For more details, → [Leo94], [Schorr74], the latter with programs.

Leakage. A term used in calorimetry to describe parts of showers which escape measurement, mostly due to their finite size, like side leakage or punchthrough (→), also uninstrumented zones. → Calorimeter.

Leakage Current. The unwanted current leaking between two electrodes under voltage. In detectors, leakage currents can be observed in wire or semiconductor detectors, without ionization caused by the passage of a charged particle. Radiation damage can increase the leakage current, which translates into a decrease of the signal-to-noise ratio.

Leakage current has, of course, a wider meaning, like when occurring with improperly grounded electrical equipment; leakage currents are also relevant in the operation of semiconductor circuits, particularly at high temperatures.

Light Attenuation. Light travelling in transparent materials [light guides (→), scintillation counters (→)] is attenuated according to an exponential law (→ Attenuation):

$$I(x,\lambda) = I(0,\lambda)\ \mathrm{e}^{-b} \quad \text{with } b = x/X_{\mathrm{A}}$$

with:

$I(x,\lambda)$ = light intensity at length x and wavelength λ

$I(0,\lambda)$ = light intensity at starting point ($x = 0$) and wavelength λ

[1] The central limit theorem states that the sum of many random variables converges towards a Gaussian distribution, if the number of variables is large, whatever the individual distribution function(s).

$X_A(\lambda)$ = attenuation length at wavelength λ.

Obviously,

$$I(X_A(\lambda)\lambda) = 1/e \; I(0, \lambda) .$$

Due to the wavelength dependence of X_A ($X_A = X_A(\lambda)$) the shape of the spectrum changes in the course of transmission. The attenuation length in scintillators can be considerably increased by the use of wavelength shifters (→).

Light Guide. Transparent material to guide a flow of light by the use of total reflection.

Typical materials are:

- plexiglass, plastic, glass;
- fibre light guides (consisting of a number of thin light guide fibres);
- liquid light guides.

Due to Liouville's law, the total area of the cross-section along a light guide cannot be reduced without light losses. For changes in direction a maximum bending (minimal bending radius) should be chosen according to the relation

$$n^2 - 1 \geq (d/2r + 1)^2$$

where

d = diameter of fibre (or light guide)

r = bending radius

n = refractive index relative to surroundings.

With a radius chosen according to the relation given above, all light entering the plane front surface of a light guide is transported due to total reflection.

Losses are due to absorption (→ light attenuation) and imperfect surfaces. Absorption and total reflection angle depend on the wavelength.

Light Yield. The light yield (or light gain) is an important parameter of scintillation counters, in particular when signal detail is relevant like when scintillators are used for sampling hadron- or electromagnetic calorimeters (→). Only a small fraction of the energy loss of a charged particle in the scintillation counter is converted into visible light. This conversion factor is usually given relative to anthracene

[Sangster56], whose light yield is of the order of 5% for blue light or about two photons/100 eV for high-energy particles. NaI has a light yield of about 230% of anthracene, while typical plastic scintillators give 50–60% [Bicron93].

For the effective light yield of scintillation counters, also the light collection, transmission and attenuation (→) play an important role. When considering the number of photoelectrons on the photocathode of a photomultiplier in a typical electromagnetic calorimeter, the conversion efficiency on the photocathode must be taken into account. As the main fraction of energy loss occurs unobserved in the absorber sheets, one ends up with typically 1000 photoelectrons per GeV energy deposit in the calorimeter (→ [Fabjan82].

Limited Streamer Tube. Sometimes simply called *streamer tube*, these are detectors based on the principle of operation in a high electric field, close to the point of breakdown (→ Gaseous Detectors, Operational Modes). Each detector element is made of a resistive cathode in the form of a (round or square) tube, with a thick (0.1 mm) anode wire in its axis. Such detectors can be produced with 1–2 cm diameter, at moderate cost, and are robust with respect to operating conditions, and hence are reliable (→ [Battistoni79], [Jonker83]).

Lorentz Angle. The angle by which particles moving in an electric field are deflected due to the effect of a magnetic field; also called *drift angle*; → Drift Chamber, Drift Velocity.

Lorentz Distribution. → Breit–Wigner Distribution

Lorentz Force. The force on a point charge q is

$$\boldsymbol{F} = q(\boldsymbol{E} + \boldsymbol{v} \times \boldsymbol{B}) \ .$$

$\boldsymbol{E} = \boldsymbol{E}(\boldsymbol{r})$ is the electric field, and $\boldsymbol{B} = \boldsymbol{B}(\boldsymbol{r})$ the magnetic induction (magnetic flux density) at the position $\boldsymbol{r}$ of the particle. c is the speed of light in vacuum, and $\boldsymbol{v} = \mathrm{d}\boldsymbol{r}/\mathrm{d}t$ is the velocity of the particle. Let

$$\begin{aligned} v &= |\boldsymbol{v}| = \mathrm{d}s/\mathrm{d}t, \\ \mathrm{d}t/\mathrm{d}\tau &= \gamma = 1/\sqrt{(1-(v^2/c^2))} \ . \end{aligned}$$

t is time, τ is proper time and s is path length. If the particle has rest mass m, its energy and momentum are

$$E = m\gamma c^2 = mc^2 \mathrm{d}t/\mathrm{d}\tau,$$

$$\boldsymbol{p} = m\gamma\boldsymbol{v} = m\ \mathrm{d}\boldsymbol{r}/\mathrm{d}\tau\ .$$

The equation of motion (neglecting bremsstrahlung (→)) is

$$\mathrm{d}\boldsymbol{p}/\mathrm{d}t = \boldsymbol{F} = q(\boldsymbol{E} + \boldsymbol{v} \times \boldsymbol{B})\ .$$

It can be written in the form

$$\mathrm{d}(\partial L/\partial \boldsymbol{v})/\mathrm{d}t - \partial L/\partial \boldsymbol{r} = 0\ ,$$

where L (the Lagrangian) is

$$L = -mc^2\sqrt{(1 - (v^2/c^2))} - q\Phi + q\boldsymbol{v} \cdot \boldsymbol{A}$$

Φ is the scalar potential and $\boldsymbol{A}$ the vector potential (→ Maxwell's equations). The canonical momentum is

$$\boldsymbol{P} = \partial L/\partial \boldsymbol{v} = \boldsymbol{p} + q\boldsymbol{A}\ ,$$

and the Hamiltonian is

$$H = \boldsymbol{p} \cdot \boldsymbol{v} - L = E + q\Phi = c((\boldsymbol{P} - q\boldsymbol{A})^2 + m^2c^2)^{1/2} + q\Phi\ .$$

The general solution of the Lorentz equation of motion contains six arbitrary integration constants. An important special case is when the fields Φ and $\boldsymbol{A}$ are time independent; then the Hamiltonian H is a constant of motion. The equation $E + q\Phi = H =$ constant can then be rewritten as

$$c\,\mathrm{d}t = \frac{\mathrm{d}s\ (H - q\Phi)}{\sqrt{(H - q\Phi)^2 - m^2c^4}}\ ,$$

and can be used to eliminate time from the remaining equations. In most experiments, time is not measured with sufficient precision to be of any interest. This means that the above equation need not be integrated, and only five integration constants are important (one of them is H). → also Equations of Motion.

Lorentz Transformation. Assume a particle of mass M at rest, $P = (M, 0)$. Under a Lorentz boost $\boldsymbol{\beta}$ it acquires four-momentum

$$P' = (E', \boldsymbol{P}') = (\gamma M, \gamma\boldsymbol{\beta}M) \quad (\text{with } \gamma = 1/\sqrt{(1 - \beta^2)})\ ,$$

i.e. after the boost it moves with a velocity $\boldsymbol{v} = c\boldsymbol{\beta}$ ($c = 1$ is the velocity of light in vacuum). The same Lorentz transformation applied to a general four-momentum $p = (e, \boldsymbol{p})$ gives the new four-momentum $p' = (e', \boldsymbol{p}')$

$$
\begin{aligned}
e' &= (eE' + \boldsymbol{p} \cdot \boldsymbol{P}')/M \\
\boldsymbol{p}' &= \boldsymbol{p} + \boldsymbol{P}'(e + e')/(E' + M) \ .
\end{aligned}
$$

The inverse transformation (replacing $\boldsymbol{P}'$ by $-\boldsymbol{P}'$) gives a four-momentum p in that frame where P is at rest. One can also separate $\boldsymbol{p}$ into components parallel and perpendicular to $\hat{\beta} = \boldsymbol{\beta}/\beta$ $(\beta \equiv |\boldsymbol{\beta}|)$,

$$
\boldsymbol{p}_L = (\boldsymbol{p} \cdot \hat{\beta})\hat{\beta} \ , \qquad \boldsymbol{p}_T = \boldsymbol{p} - \boldsymbol{p}_L \ .
$$

Then:

$$
\begin{aligned}
\boldsymbol{p}'_L &= \gamma(\boldsymbol{p}_L + \boldsymbol{\beta} e) \ , \\
\boldsymbol{p}'_T &= \boldsymbol{p}_T \ , \\
e' &= \gamma(e + \beta|\boldsymbol{p}_L|) = \gamma(e + \boldsymbol{\beta} \cdot \boldsymbol{p}) \ .
\end{aligned}
$$

We have defined here *active* Lorentz transformations, i.e. four-vectors are transformed, but the reference system is not transformed. The *passive* point of view is that a four-vector is not transformed, but its components are, because they are given relative to a transformed coordinate system. Let u_0, u_1, u_2, u_3 be the original unit four-vectors, and let the transformation be

$$
u'_n = L u_n = L_n^m u_m \ .
$$

Then

$$
p = p^m u_m = p'^n u'_n = p'^n \ L_n^m \ u_m \ ,
$$

so that the relation between the new components p'^m and the old p^m is

$$
\begin{aligned}
p^m &= L_n^m p'^n \ , \\
p'^m &= (L^{-1})_n^m p^n \ .
\end{aligned}
$$

Now assume a particle of mass M with energy E and three-momentum $\boldsymbol{P}$ with respect to reference system I, let system II be the rest system of the particle and let p be an arbitrary four-vector in system I. Then

In system I:

$$
\begin{aligned}
P &= (E, \boldsymbol{p}) \\
p &= (e, \boldsymbol{p}) \ ,
\end{aligned}
$$

In system II:

$$\begin{aligned} P &= (M, 0) \\ p &= (e', \boldsymbol{p}') \\ e' &= (eE - \boldsymbol{p} \cdot \boldsymbol{P})/M \\ \boldsymbol{p}' &= \boldsymbol{p} - \boldsymbol{P}(e + e')/(E + M) \,. \end{aligned}$$

The clear distinction between active and passive transformations is important, in particular when two or more successive transformations are to be performed. For a more detailed discussion of this point → [Bock98].

Luminosity. The luminosity L defines the intensity of colliding beam machines. Luminosity is an important parameter when deriving cross-sections from events measured over a period of time: the count in a given class of events is given by

$$N_{class} = A\, \sigma_{\text{class}} \int L\, \mathrm{d}t \,,$$

where A is the acceptance (→), σ the cross-section (→), and $\int L\, \mathrm{d}t$ the luminosity integrated over time.

Luminosity is defined by the accelerator and beam parameters; if the two beams are continuously distributed around a ring, and traverse each other under a (small) angle α, then

$$L = N_1 N_2 f / (2\pi R\, h\, \tan \alpha/2) \,,$$

where N_1 and N_2 are the number of particles in the two beams, f is the frequency of rotation, $2\pi R$ is the ring circumference and h is the beam width perpendicular to the ring, i.e. in the vertical direction (note that the beam shape in the plane of the ring does not enter the formula, as in that projection all particles cross the path of all counterrotating particles, whatever the beam spread).

For beams colliding in bunches and head-on, the luminosity is given by

$$L = N_1 N_2 f W_x W_y \,,$$

where N_1 and N_2 are the number of particles in bunch 1 and 2, and f is the frequency of rotation. W_x, W_y are defined from beam profiles by

$$\begin{aligned} W_x &= \int D_{1x}(x)\, D_{2x}(x)\, \mathrm{d}x \Big/ \left(\int D_{1x}(x)\, \mathrm{d}x \int D_{2x}(x)\, \mathrm{d}x \right) \\ W_y &= \int D_{1y}(y)\, D_{2y}(y)\, \mathrm{d}y \Big/ \left(\int D_{1y}(y)\, \mathrm{d}y \int D_{2y}(y)\, \mathrm{d}y \right) \,, \end{aligned}$$

where $D(x)$, $D(y)$ are the independent particle densities in the horizontal and vertical directions, subscripts 1, 2 refer to the two bunches and the integrals extend over the full beam size.

If the beam particle densities are assumed Gaussian, along the same axis, and the same in x and y, one can set

$$W_x W_y = 2\pi(\sigma_1^2 + \sigma_2^2) ,$$

where σ^2 is the variance. Accelerator physicists usually transform the formula into different variables, using ε_i, the characteristic (invariant) beam emittance given in [mrad], and β_i, the transverse betatron amplitude; they relate to the above by the formula $\varepsilon_i \beta_i = \pi\sigma_i^2$. Frequently, one also replaces the number of particles by the beam current I, given by $I = Nfe$, with e the elementary charge (assuming singly charged beam particles, of course).

Luminosities typical for some colliding beam machines (for more details, → [Barnett96]) are:

Machine	Nominal energy [GeV]		Luminosity (max.) [$\mathrm{cm^{-2}s^{-1}}$]
SLC (SLAC)	50 + 50	e^+e^-	8×10^{29}
TRISTAN (KEK)	32 + 32	e^+e^-	4×10^{31}
VEPP4 (Novosibirsk)	6 + 6	e^+e^-	5×10^{31}
LEP (CERN, Geneva)	90 + 90	e^+e^-	3×10^{31}
HERA (DESY, Hamburg)	30 + 820	e^-p	1.6×10^{31}
Tevatron (FNAL, Batavia)	1000 + 1000	$p\bar{p}$	2.5×10^{31}
LHC (CERN, under design)	7000 + 7000	pp	3×10^{34}

For a more detailed discussion and a good introduction to accelerator physics, → [Bryant93] or [Scharf86]. For more detailed reading, the CERN Accelerator School has produced invaluable reading material during its 15 years of existence (e.g. [CAS96]).

Mandelstam Variables. Mandelstam variables are Lorentz-invariant variables describing the kinematics of particle reactions. Originally the variables were introduced by Mandelstam [Mandelstam58] to describe two-body elastic scattering amplitudes in terms of dispersion relations as functions of two complex variables s and t. Mandelstam variables are also widely used now to describe the kinematics of multibody final states viewed as two incident and two outgoing systems.

Calling the incident particles 1 and 2, the outgoing systems 3 and 4 and using four-momentum conservation

$$p_1 + p_2 - p_3 - p_4 = 0 \quad \text{and} \quad (p_1)^2 = m_i^2$$

one can write the Mandelstam variables as follows

$$\begin{aligned} s &= (p_1 + p_2)^2 = (p_3 + p_4)^2 \\ t &= (p_1 + p_3)^2 = (p_2 + p_4)^2 \\ u &= (p_1 + p_4)^2 = (p_2 + p_3)^2 \end{aligned}$$

from which it follows that

$$s + t + u = \text{const.} = m_1^2 + m_2^2 + m_3^2 + m_4^2 .$$

The usual assignment of the indices $1, \ldots, 4$ is such that 3 appears as the system produced most naturally (by quantum numbers) from 1 (e.g. if 1 is the projectile, 3 is the scattered system); t then peaks at values close to zero.

From the above definitions it follows that s is equivalent to the square of the centre-of-mass energy of the reaction, and t and u correspond to the square of the four-momentum transfer in the direct and exchange channel respectively.

For a given s, both t and u depend linearly on the cosine of the centre-of-mass deflection angle by

$$\begin{aligned} -t &= 2E_1^* E_3^* - m_1^2 - m_3^2 - 2p_1^* p_3^* \cos\Theta^* \\ -u &= 2E_2^* E_3^* - m_2^2 - m_3^2 - 2p_2^* p_3^* \cos\Theta^* . \end{aligned}$$

In the case of elastic scattering and again for fixed s, t is given by

$$t = -2p^{*2}(1 - \cos\Theta^*)$$

and has the bounds

$$0 \leq -t \leq 4p^{*2} .$$

As $s + t + u = \text{const.}$, a natural representation of a reaction in terms of Mandelstam variables is a diagram having s, t, u as axes at 120°. The physical boundary is then a third-order algebraic curve, with the three axes acting as asymptotes. When used for decays, s, t and u are the squares of effective masses, and the physical region is a closed area. → also Dalitz Plot.

Maxwell's Equations. Maxwell's equations (in macroscopic form and MKSA units) are

$$
\begin{aligned}
\nabla \cdot \boldsymbol{B} &= 0 \, , \\
\nabla \times \boldsymbol{E} + \partial \boldsymbol{B}/\partial t &= 0 \, , \\
\nabla \cdot \boldsymbol{D} &= \rho \, , \\
\nabla \times \boldsymbol{H} - \partial \boldsymbol{D}/\partial t &= \boldsymbol{J} \, .
\end{aligned}
$$

(Other system units are discussed e.g. in [Jackson75].)

$\boldsymbol{E}$ is the electric field, $\boldsymbol{D} = \varepsilon_0 \boldsymbol{E} + \boldsymbol{P}$ is the displacement, ε_0 the permittivity of free space and $\boldsymbol{P}$ the polarization. $\boldsymbol{B}$ is the magnetic induction (magnetic flux density), $\boldsymbol{H} = \boldsymbol{B}/\mu_0 - \boldsymbol{M}$ the magnetic field, μ_0 the permeability of free space and $\boldsymbol{M}$ the magnetization. ρ is the density of electric charge and $\boldsymbol{J}$ is the current density. The relations between $\boldsymbol{E}$ and $\boldsymbol{D}$, and between $\boldsymbol{B}$ and $\boldsymbol{H}$, are called constitutive equations; they describe the medium. In a linear, isotropic medium $\boldsymbol{D} = \varepsilon \boldsymbol{E}$ and $\boldsymbol{H} = \mu \boldsymbol{B}$, where ε and μ are constants. In general $\boldsymbol{H}$ (or $\boldsymbol{D}$) is not even a unique function of $\boldsymbol{B}$ (or $\boldsymbol{E}$), but depends upon the earlier time evolution (hysteresis).

$$\varepsilon_0 \mu_0 = c^{-2}$$

where c is the speed of light in vacuum and by definition

$$\mu_0 = 4\pi \; 10^{-7} \; NA^{-2} \, .$$

The continuity equation

$$\partial \rho / \partial t + \nabla \cdot \boldsymbol{J} = 0$$

follows from Maxwell's equations and expresses the conservation of electric charge.

Defining the scalar potential Φ and vector potential $\boldsymbol{A}$ by

$$
\begin{aligned}
\boldsymbol{E} &= -\nabla \Phi - \partial \boldsymbol{A}/\partial t \, , \\
\boldsymbol{B} &= \nabla \times \boldsymbol{A} \, ,
\end{aligned}
$$

explicitly solves half of Maxwell's equations. The potentials are not unique, since any gauge transformation

$$\Phi, \quad \boldsymbol{A} \rightarrow \Phi + \partial \chi / \partial t, \quad \boldsymbol{A} - \nabla \chi$$

leaves the physical fields $\boldsymbol{E}$ and $\boldsymbol{B}$ unchanged, χ being an arbitrary function.

Other types of potentials may be useful in certain cases (e.g. a scalar potential for $\boldsymbol{B}$).

Mean Free Path. The mean free path of a particle in a medium is a measure of its probability of undergoing interactions of a given kind. It is related to the cross-section corresponding to this type of interaction by the formula

$$\sigma\lambda = \Omega/N = A/(N_A\rho) ,$$

with

$$\begin{aligned}
\sigma &= \text{cross} - \text{section } [\text{cm}^2] \\
\lambda &= \text{mean free path } [\text{cm}] \\
\Omega &= \text{volume of interaction} \\
N &= \text{Number of target particles in } \Omega \\
A &= \text{atomic weight } [\text{g/mole}] \\
N_\text{A} &= \text{Avogadro's number } (6.022\ 10^{23}/\text{mole}) \\
\rho &= \text{density } [\text{g/cm}^3]
\end{aligned}$$

The mean free path is the average of a distribution of distances following an exponential law:

$$P(l)\,\mathrm{d}l = \lambda^{-1}\ \mathrm{e}^{-1/\lambda}\,\mathrm{d}l\ .$$

Tables often give the quantity $\lambda\rho = A/(\sigma N_\text{A})$ (in g/cm^2) instead of λ (in cm). For some numerical values, → Collision Length.

Microstrip Detector. A detector made of a large number of identical detector structures usually in a plane (a fragmented metal electrode on a common support, an insulator, or a semiconductor), in order to obtain a device with high resolution in one dimension, in that plane. Two main types of microstrip detectors are common: silicon microstrip detectors [Peisert92] and microstrip gas chambers (→).

The main geometrical parameter of a microstrip structure is the pitch (→); typical pitch values for silicon microstrip detectors are in the range of ten to one hundred micrometres.

Microstrip detectors can be built as single- or double-sided devices. The readout is normally done in channels connecting several strips.

For more details, → Semiconductor Detectors.

Microstrip Gas Chamber. The microstrip gas chamber (MSGC) is a high-precision and high-rate tracking detector, for high-energy physics applications. It was introduced by Oed in 1988 [Oed88], and later optimized for tracking at high energies [Angelini91].

MSGC's more or less reproduce the field structure of multiwire chambers (MWPC); they are made of a sequence of alternating thin anode and cathode strips on an insulating support; a drift electrode on the back plane defines a region of charge collection, and application of appropriate potentials on the strip electrodes creates a proportional gas multiplication field.

The classical MSGC is built on a glass support of thickness of the order of a few hundred μm, and the drift volume is defined by a drift cathode situated at a typical distance of 2–6 mm from the plane of the strips. The typical pitch (the repetition sequence) is 100–200 μm. The anodes and cathodes are deposited on the support using techniques from microelectronics, e.g. planar technology.

Other constructive variants use semiconductor supports, silicon oxide, implanted special conductive glasses, quartz, or plastics (kapton, tedlar, upilex). A major research and development effort has been invested in optimizing designs (→ [Bouclier92]).

The performances achieved by these detectors are:

- energy resolution: FWHM of 11–18% for the 6 keV X-ray emitted by ^{55}Fe;
- intrinsic spatial resolution: 30 μm rms using the method of centre of gravity of the amplitude pulses;
- multitrack resolution of about 250 μm.

The technical solutions on support and filling gas mixtures are still under development, → Gas Mixtures in Gaseous Detectors, or e.g. [DellaMea94].

Minimum Ionization. The ionization energy loss (→) of charged particles in a medium is primarily a function of its velocity β. It reaches a minimum at values of $\beta\gamma$ between 3 and 4. The minimum ionization is of the order of a few thousand ionizations per $(\mathrm{g\,cm^{-2}})$ in most materials, quite independent of the particle mass. The abbreviation *mip* is often used for *minimum-ionizing particle.*

Missing Mass. The missing mass is the effective mass (→) of all particles missing in a partly measured interaction. Separating into particles of initial and final state, the missing mass is defined by

$$m_{\mathrm{miss}}^2 = \left(\sum E_{\mathrm{unit}} - \sum E_{\mathrm{final}}\right)^2 - \left(\sum \mathbf{p}_{\mathrm{init}} - \sum \mathbf{p}_{\mathrm{final}}\right)^2 .$$

In exclusive measurements, m_{miss} can serve as a test for the completeness of measurement: no missing particle results in $m^2_{\text{miss}} = 0$.

The observed value of m^2_{miss} may well be negative due to measurement errors, whereas an effective mass computed from observed particles will always be real, even in the presence of measurement errors. Note that the error propagation for the case $m_{\text{miss}} = 0$ gives in first order a zero error for m^2_{miss}, and higher order terms result in a strongly asymmetric probability distribution for m^2_{miss} peaking at negative values.

Moliere's Formula. Approximates the projected scattering angle of multiple scattering (→) by a Gaussian, with a width

$$\theta^2_{\text{proj}} = \left(\frac{0.015}{\beta P}\right)^2 \frac{z^2 x}{X_0} \left[1 + 0.12 \log_{10}(x/X_0)\right]^2,$$

with β, P, and z the particle velocity, momentum (in GeV/c) and charge, x the depth of the traversed material, and X_0 the radiation length. For a more detailed discussion and further approximations, → [Barnett96].

Moliere Radius. A characteristic constant of a material describing its electromagnetic interaction properties, and related to the radiation length (→) by

$$R_{\text{M}} = 0.0265 X_0 (Z + 1.2)$$

with X_0 the radiation length and Z the atomic number. R_{M} is a good scaling variable in describing the transverse dimension of electromagnetic showers (→). For more discussion, → also [Barnett96].

Möller Scattering. Scattering of two electrons or positrons in each other's field ($e^- e^-$ or $e^+ e^+$). For $p \gg m_e c$ one obtains in first-order perturbation theory for the differential cross-section in the centre-of-mass system (cms):

$$\mathrm{d}\sigma/\mathrm{d}\Omega(e^- e^- \to e^- e^-) = (r_{\text{e}}^2/4)(m_{\text{e}} c/p)^2 (3 + \cos^2\theta)^2 / \sin^4\theta$$

with

$$\begin{aligned} r_{\text{e}} &= \text{classical electron radius} \\ m_{\text{e}} &= \text{rest mass of electron} \\ p &= \text{momentum in the cms} \end{aligned}$$

Momentum Flow. A presentation of the momentum vectors in a reaction, or more frequently of the average momentum vectors, by binning them in solid angle variables. This technique has been used for momenta as measured in track devices, or for energy deposition in calorimeter cells, often as an analysis method in jet studies. In some experiments (e^+e^- collisions), events have been described entirely in terms of the *momentum tensor*: the eigenvalue analysis (*principal components*) of the matrix $A_{i,j} = \sum_{\text{tracks}} p_i p_j$ gives a reaction-specific ellipsoid (the p_i are momentum components along axis i, and the pair i, j goes through all permutations of coordinate axes).

Momentum Kick. → Equations of Motion

MSGC. Short for Microstrip Gas Chamber (→).

Multiple Scattering. Effect of Coulomb scattering (→) acting on a particle and summing up in the way of many relatively small random changes of the direction of flight. For a thin layer of traversed material the variance of the projected scattering angle of a particle with unit charge can be approximated by

$$\begin{aligned}\langle \Theta^2_{\text{proj}} \rangle &= \langle \Theta^2_{\text{space}} \rangle / 2 \\ &= \frac{(21\ \text{MeV})^2 (m^2 + p^2)}{2p^4 \beta^2} \, \frac{x}{X_0} \left(1 + 0.038 \log \frac{x}{X_0}\right)^2\end{aligned}$$

where

$\langle \Theta^2 \rangle$ = expectation value of the square of the scattering angle per unit length

m = mass of the (heavy) projectile

p = momentum of the (heavy) projectile

β = velocity of the (heavy) projectile

X_0 = radiation length (→) of the traversed material

x = thickness of the traversed material.

The underlying assumption of a Gaussian distribution makes this approximation a crude one; in particular, large angles are underestimated by the Gaussian form. For more details → [Rossi65], [Scott63], [Fernow86], [Barnett96].

In the general case, the scattering effect, considered as white noise, is described by

$$\varepsilon_K = \int \frac{\partial C_k}{\partial \Theta}\,\mathrm{d}\Theta = \int_0^{S_k} \frac{\partial C_k}{\partial \Theta(s)}\theta(s)\,\mathrm{d}s, \qquad \Theta(s) = \int_0^{s} \theta(s)\,\mathrm{d}s\ ,$$

with s = path length, $\partial C_k/\partial\Theta(s)$ = influence of the scattering angle θ at s on the impact point C in detector k, and $\theta(s) = \partial\Theta/\partial s$ = white noise.

For a straight track in a homogeneous medium and with detectors perpendicular to the track ($x \equiv s$, $y(0) = 0$, $(\mathrm{d}y/\mathrm{d}x)_0 = 0$), it follows that

$$E(y^2) = E\left(\int_0^x \int_0^x (x-x')(x-x'')\theta(x')\theta(x'')\,\mathrm{d}x'\mathrm{d}x''\right)\ ,$$

and with (writing Θ_s^2 for Θ_space^2) $E[\theta(x')\theta(x'')\,\mathrm{d}x''] = \Theta_\mathrm{s}^2\delta(x'-x'')\,\mathrm{d}x''/2$:

$$E(y^2) = (\Theta_\mathrm{s}^2/2)\int (x-x')^2\,\mathrm{d}x' = (\Theta_\mathrm{s}^2/2)(x^3/3)\ ,$$

and similarly

$$\begin{aligned} E(\Theta_y^2) &= (\Theta_\mathrm{s}^2/2)x \\ E(y,\Theta_y) &= (\Theta_\mathrm{s}^2/2)(x^2/2)\ , \end{aligned}$$

or written as a matrix

$$\mathrm{cov}(y,\Theta_y;x) = (\Theta_s^2/2)\begin{pmatrix} x^3/3 & x^2/2 \\ x^2/2 & x \end{pmatrix}\ .$$

Up to quadratic properties this is equivalent to the Gaussian probability density function

$$\mathrm{d}(y,\Theta_y;x) = \sqrt{2\pi x^4\Theta_\mathrm{s}^2/24}\,\mathrm{e}^{-(4/\Theta^2)(\Theta_y^2/x)-(3y\Theta_y/x^2)+(3y^2/x^3)}\ .$$

The effects of multiple scattering on track reconstruction were first described by Gluckstern [Gluckstern63]. In track fitting a matrix formalism for multiple scattering can be used. To the (usually diagonal) covariance matrix describing the detector resolution (→) a non-diagonal term taking into account multiple scattering must be added:

$$\begin{pmatrix} \sigma_1'^2 & .. & \sigma_1'\sigma_n' \\ .. & .. & .. \\ \sigma_n'\sigma_1' & .. & \sigma_n'^2 \end{pmatrix} = \begin{pmatrix} \sigma_1^2 & .. & 0 \\ .. & .. & .. \\ 0 & .. & \sigma_n^2 \end{pmatrix} + \begin{pmatrix} E(\varepsilon_1^2) & .. & E(\varepsilon_1\varepsilon_n) \\ .. & .. & .. \\ E(\varepsilon_n\varepsilon_1) & .. & E(\varepsilon_n^2) \end{pmatrix}$$

where ε_i is a random variable describing the change of the ith measurement due to multiple scattering for particles travelling parallel to the x-axis and detectors normal to this axis, and E stands for expectation value.

For discrete scatterers (obstacles) and particles moving parallel to the x-axis and detectors normal to this axis, this covariance matrix is given by

$$E(\varepsilon_j \varepsilon_k) = \sum (x_j - x_i)(x_k - x_i) E(\Theta_i^2) \ ,$$

The sum is over all obstacles with $x_i < \min(x_j, x_k)$.

A detailed discussion of this matrix formalism is given in [Eichinger81].

Multiwire Chamber. A detector for charged particles which essentially consists of thin parallel and equally spaced anode wires symmetrically sandwiched between two cathode planes. Cathode planes can be a set of thin equally spaced wires but also can be made of a continuous plane conductor. The gap between the plane of the anode wires and the cathode plane is normally a few (3 to 4) times the spacing between the anode wires. The cathodes are on negative voltage and the wires are grounded. This creates a homogeneous electric field in most regions, with all field lines leading from the cathode to the anode wires. Around the anode wires, the field increases rapidly. If a particle passes through the detector it ionizes the gas (→ Gas Mixtures in Gaseous Detectors) in the chamber, and the liberated electrons follow the electric field lines towards the anode wires. The strong field very close to the wire acts as a *multiplication region*: the energy of the electrons increases, and in turn they ionize the gas, causing an avalanche of electrons to reach the anode wire.

The principles underlying modern multiwire chambers were already shown around 1920 (Geiger–Müller counter); the first wire chamber used in high-energy physics was, in fact, a spark chamber, whose electrode plates were replaced by grids of parallel wires in order to reduce multiple scattering, energy loss and secondary interactions, and to allow the localization of particle impact points without using photographic methods. Later, the idea of the Geiger–Müller counter was taken up again and developed into modern position detectors, mostly by the work of G. Charpak and his coworkers (Nobel prize 1992), [Charpak68].

The pulses are read from the anode wire or *sense wire*. The pulse height depends on the gas used and the voltage applied and also geometrical parameters of the chamber like the gap, wire spacing, wire diameter, etc. If the chamber is used in proportional mode (→ Gaseous Detectors, Operational Modes), the pulse height is a measure of the energy loss of the particle in the gas. This can be used for particle or momentum identification. Simple multiwire chambers are used as tracking chambers, with the anode wires only giving one bit of information for a passing particle. Multiple planes with different angles of inclination for the wires will then allow reconstruction of trajectories in space.

A wide variety of multiwire chambers of different complexity has been constructed and tested, and used successfully in experiments:

- proportional and drift chambers of planar and cylindrical type. They provide one-dimensional measurements on a surface made by the parallel wires. The information is binary (on or off) or may contain pulse height; with drift time measured, the inter-wire distance can be subdivided, but left-right ambiguities are introduced. Planes at different angles or segmentation of the cathode (→ Cathode Strips) allow one to obtain information along the wire.
- jet and time projection chambers. These chambers are characterized by comparatively longer drift times; multiple sense wires can cover a large volume, and are usually equipped with multihit electronics, such that the passage of several tracks in the volume part associated with a wire can be recorded. These chambers give inherently two-dimensional information (the drift time is in a coordinate orthogonal to the wire plane). With added cathode instrumentation or charge division, three-dimensional points can be obtained. For more details, → Drift Chamber; a good review can be found in [Blum93].
- time expansion chambers (→), a special type of drift chamber.

The main parameters of a wire chamber (from the viewpoint of optimizing particle detection) are:

- single- and multihit detection efficiency;
- precisison and two-track separation;
- dead time.

For more details, → [Sauli91], [Fabjan80], [Walenta71], or the proceedings of the Vienna wire chamber conferences (→ e.g. [Krammer95]).

MWPC. Short for Multiwire Proportional Chamber (→ Multiwire Chamber).

Neutron Fluence. Particle fluence is defined as the number of particles traversing a unit area in a certain point in space in a unit period of time. Most frequently, it is measured in cm^{-2}.

In particular, neutron fluence in high-energy physics applications is of interest in the context of the radiation environment around the interaction regions of colliders; it serves as a measure for potential radiation damage for the detector systems to be used there. It is common practice to express charged and neutral particle contributions to radiation in terms of dose (→ Radiation Measures and Units) and 1 MeV neutron equivalent fluence (→ also NIEL Scaling), respectively.

The *1 MeV equivalent neutron fluence* is the fluence of 1 MeV neutrons producing the same damage in a detector material as induced by an arbitrary particle fluence with a specific energy distribution. The choice of this particular normalization is partly due to historical reasons, as the standard energy to scale to was considered first in damage studies in the MeV range, in neutron physics; however, there is also a physical background: the neutron spectra expected in detectors at future hadron colliders typically have a probability density peaking in this energy region.

NIEL Scaling. According to NIEL (non-ionizing energy loss) scaling, any particle fluence can be reduced to an equivalent 1 MeV neutron fluence producing the same bulk damage in a specific semiconductor. The scaling is based on the hypothesis that generation of bulk damage is due to non-ionizing energy transfers to the lattice.

Given an arbitrary particle field with a spectral distribution $\varphi(E)$ and of fluence Φ, the 1 MeV equivalent neutron fluence is:

$$\Phi_{\mathrm{eq}}^{1\,\mathrm{MeV}} = \kappa\Phi\,.$$

κ is called the *hardness parameter* and is defined as:

$$\kappa = \frac{\mathrm{EDK}}{\mathrm{EDK}(1\,\mathrm{MeV})}$$

with EDK the energy spectrum averaged displacement KERMA (→ Radiation Measures and Units):

$$\mathrm{EDK} = \frac{\int D(E)\varphi(E)\,\mathrm{d}E}{\int \varphi(E)\,\mathrm{d}E}$$

where $\varphi(E)$ is the differential flux, and

$$D(E) = \sum_k \sigma_k(E) \int \mathrm{d}E_{\mathrm{R}} f_k(E, E_{\mathrm{R}}) P(E_{\mathrm{R}})$$

is the displacement KERMA or the *damage function* for the energy E of the incident particle, σ_k the cross-section for reaction k, $f_k(E, E_{\mathrm{R}})$ the probability of the incident particle to produce a recoil of energy E_{R} in reaction k, and $P(E_{\mathrm{R}})$ the *partition function* (the part of the recoil energy deposited in displacements). EDK (1 MeV) = 95 MeV mb [ASTM93]. The integration is done over the whole energy range.

A few damage functions are available for neutrons up to 18 MeV in silicon (→ [Lazo86], [Ougouag90] and have been standardized in [ASTM93].

[Ougouag90] also gives displacement function tables for GaAs.

A review and analysis for neutrons can be found in [Vasilescu97] (→ also [Angelescu96]).

For neutrons above 18 MeV and for other particles, the situation is still controversial. Results on neutrons are available from [Ginneken89], who also studied electron, muon, pion, gamma and proton NIEL, and [Konobeyev92] and [Huhtinen93b]. For protons, the energy region up to 200 MeV is covered by [Summers93] (also giving tabulations for GaAs and InP, for protons and electrons). Above 200 MeV the only extrapolations are those from [Huhtinen93b] and [Ginneken89]. For pions, damage function calculations are presented in [Ginneken89], [Huhtinen93b], [Lazanu97]. These results should be treated with some care. New calculations are needed, based on detailed simulation and comparison to the experiment, as in [Huhtinen93a], [Aarnio95]. Work is in progress in further analysis of the NIEL, due to its importance especially for damage estimates in collider radiation environments and the operational scenario of the 10 years of experiments planned at the LHC at CERN.

Noise. Random background signals, mostly in transmission or communication systems. Noise is strictly dependent on the systems used and their technologies. One usually distinguishes *white noise* which occurs with similar amplitudes over a wide frequency spectrum (the analogy is with white light, made up of all visible frequencies), and is also called random, Gaussian or steady state noise, and *impulse noise* which is a momentary disturbance, limited in the frequency band. In analogue electronics, one talks about *shot noise*; this is Poisson-distributed and is explained by the small statistics of charge carriers passing through semiconductor junctions; in image processing, the expression *blue noise* is used for random perturbations favouring high over low frequencies (sometimes also called $1/f$ *noise*, where f is the frequency).

In experiments, noise is often and quite generally (and imprecisely) used as a synonym for backgrounds of different kinds; outliers (→) are noise of the impulse type; multiple scattering of particles produces fluctuations of the white noise type.

Nuclear Emulsion. An emulsion is made, as for photographic film, of a silver salt, usually bromide, embedded in gelatine and spread thinly on a substrate. Multiple layers of emulsion were historically the first means of visualizing charged particle tracks; emulsion stacks are still used today to record, with very high positional precision, very short tracks (e.g. tau leptons, which have a track length of less than a millimetre), or in other circumstances demanding very high precision.

Emulsions are permanently sensitive and cause nontrivial data acquisition work by microscopic methods; usually, emulsions are left in place for long runs, and hence are restricted to applications in areas of small particle flux or in low-cross-section experiments, like neutrino physics. Data acquisition by automated means (e.g. by scanning the film with a CCD camera) has been found possible in some circumstances.

Occupancy. A term used for the average probability of a single detector cell (like a wire in a drift chamber, a strip in a microstrip detector, a cell in a calorimeter) to be hit by at least one particle, in a given interaction or unit of time.

Optical Theorem. The optical theorem relates the forward elastic scattering amplitude to the total cross-section for the same particles by the relation:

$$\mathrm{Im}\, f(0) = k\sigma_T ,$$

where $f(\Theta)$ is the elastic scattering amplitude at centre-of-mass scattering angle Θ, $k = q/4\pi$, is a proportionality constant, q is the centre-of-mass momentum, and σ_T is the total cross-section. The name optical theorem derives from the original application to electromagnetic phenomena in optical media.

By $d\sigma_{el}/d\Omega(\Theta = 0) = |f(0)|^2 = (1 + \rho^2)k^2\sigma_T^2$, (where ρ is the ratio of the real to the imaginary parts of the scattering amplitude at $\Theta = 0$), the optical theorem allows one to use measurements of elastic differential cross-sections at very small angles (by extrapolation to $\Theta = 0$) for the determination of the total cross-section.

Outlier. The statistical term for something physicists often also call "noise". An outlier is an observation which does not correspond to the phenomenon being studied, but instead has its origin in background or in a gross measurement (or assignment) error. In practice, nearly all experimental data samples are subject to contamination from outliers, a fact which reduces the real efficiency of theoretically optimal statistical methods.

Pad. Pads are rectangular or square conductors used as readout cathodes in tracking chambers and in calorimetry, for localizing tracks or showers (→ Calorimeter, also → [Fretwurst96], [Donaldson89]).

The name also refers to a specific geometry of semiconductor detectors containing one single diode, with surface dimensions ranging from a few mm to a few cm (compare to microstrips, where the width is in the micron range, and the length is typically several cm).

Pads in semiconductor detectors may be circular or rectangular. Very small-area pads can be assimilated to pixels (pixel detectors are occasionally called pad detectors). Large pads are sometimes called tiles [Heijne89].

Pair Production. The dominant interaction process for high-energy photons. Only at low energies (below 10 MeV) do Compton scattering and photoelectric absorption compete in cross-section.

For the calculation of electromagnetic showers (→), the energy spectrum of the generated positron (electron) can be approximated by [Lohrmann81]:

$$\Phi(E_+, k)\,dE_+dx \approx (dx/X_0)(dE_+/k)[\nu^2 + (1 - \nu^2) + (2/3)\nu(\nu - 1)]\,,$$

with

X_0 = radiation length (→)
E_+ = energy of the produced positron
k = energy of the incident photon

$$\mathrm{d}x = \text{thickness of the traversed material}$$
$$\nu = E_+/k \ .$$

This formula holds for $E_+ \gg m_e \cdot Z^{-1/3}$, and for k >10 MeV, where

$$m_e = \text{electron mass}$$
$$Z = \text{atomic number of the traversed material.}$$

The total probability for pair production over a path $\mathrm{d}x$ is given by

$$\int_{m_e}^{k} \Phi(E_+, k)\,\mathrm{d}E_+ \approx (7/9)\,\mathrm{d}x/X_0\,;$$

expressed more simply, the attenuation length due to pair production is 9/7 times the radiation length.

Partial Wave Analysis. A technique to extract detailed scattering amplitudes from experimental data, usually restricted to comparatively low energies. The partial wave formalism expands particle wave functions in terms of complex amplitudes, which are defined for given spin and parity. The amplitudes are parameterized, often in terms of spherical harmonics. Typically, the information available from measurements is insufficient to obtain unambiguous results, and additional physics assumptions have to be introduced. Since these depend on the reaction under study, the approximations and limitations cannot easily be generalized. → [Hamilton72], [Litchfield84]. For a recent analysis of a world-wide collection of nucleon–nucleon reactions at low energies (stored in a database), → [Stoks93].

Partial Width. This term is used as a synonym for the partial cross-section of one of several competing reactions, i.e. is directly proportional to a branching ratio. The term comes from the level widths of excited nuclei with different lifetimes τ_i. Lifetime and decay width Γ are related by the Heisenberg uncertainty principle

$$\Gamma_i = \hbar/\tau_i \ ,$$

where $\hbar = h/2\pi = 6.58\ 10^{-25}$ for τ in s and Γ in GeV. The total lifetime is then given by

$$1/\tau = \sum(1/\tau_i) \ ,$$

or the total width by

$$\Gamma = \sum \Gamma_i \ .$$

An individual relative branching ratio is given by Γ_i/Γ.

Particle Identification. Certain detectors have as their main objective the identification of particles by their mass or quantum numbers, as opposed to position-sensitive detectors used for tracking, or calorimeters used for measuring particle energy. Particle identification relies on special properties of some particles, like muons which carry charge but do not shower nor interact strongly, or the electromagnetic shower characteristic for electrons and gammas. In other cases, the mass sensitivity of some radiation can be used (Cherenkov or transition radiation), or the mass dependence of ionization loss ($\mathrm{d}E/\mathrm{d}x$).

Many detectors, of course, have combined functions: tracking detectors can also be used to sample ionization (e.g. [Breuker87]), or they may be built with interspersed layers provoking transition radiation; muon detectors, by their shielding (typically through a calorimeter), identify muons simply by the fact that they have not been absorbed; calorimeters absorb electrons in a way different from hadrons, and can also produce a useable signal for a single minimum-ionizing particle, measuring particle showers at the same time.

→ Cherenkov Counter, → Transition Radiation, → Ionization Sampling, → Calorimeter. For a detailed treatment, → [Allison91].

Peyrou Plot. The Peyrou plot is a scatter plot showing longitudinal centre-of-mass momentum versus transverse momentum; a plot useful for inclusively measured single particles.

Photodiodes. Photodetectors based on semiconductor technology. Photodiodes make use of the photovoltaic effect: the generation of a voltage across a p-n junction of a semiconductor, when the junction is exposed to light. The term is broadly used, including even solar elements; it usually refers to sensors used to measure the intensity of light. In high-energy physics they are used as readout elements associated to scintillators, e.g. for scintillating fibres and for some calorimeters (→ e.g. [Fenker91]).

Their main features are:

- excellent linearity and low noise (limited by shot noise);

- wide spectral response and high quantum efficiency (e.g. about 80% at 800 nm);
- easy calibration;
- insensivity to magnetic fields;
- compactness and mechanical ruggedness;
- stability and long life.

Photodiodes are very similar to rectifying junction diodes, and are manufactured as p-n Si or GaAsP photodiodes, PIN Si photodiodes, GaAsP Schottky and Si avalanche photodiodes. The p layer is at the light sensitive surface, and the n-side at the substrate. An avalanche photodiode is obtained by adding to a simple PIN diode an electron multiplication region, viz. an area with large bias voltage generating secondary electrons and holes. This process multiplies the signal (and primary shot noise), and adds stochastic noise for the multiplication process; they are not suited for single-photon readout. Depending on the bias voltage, a proportional mode (allowing one to measure light intensity) or a Geiger mode (giving a larger, saturated signal) can be achieved. An application for scintillating fibre tracking has been discussed in [Nonaka96].

For an overview, → [Kazovsky96]. The radiation hardness of photodiodes has been studied in parallel with semiconductor detectors (→ Radiation Damage in Semiconductors, or [Hall90]).

Photoelectric Effect. One of the processes contributing to the attenuation of γ's in matter, together with Compton scattering and pair production. The photoelectric cross-section rises with Z^5 (Z being the atomic number of the medium), but is proportional to $1/E^3$, hence plays a role only at low energies (say at $E \leq 100$ keV).

Photomultiplier. A device to convert light into an electric signal (the name is often abbreviated to PM). Photomultipliers are of great relevance in all detectors based on scintillating material. A photomultiplier consists of a photocathode (photons are converted into electrons, making use of the photoelectric effect), a multiplier chain (strings of successive electron absorbers with enhanced secondary emission, called *dynodes*, the entire string using electric fields to accelerate electrons), and an anode, which collects the resulting current. Commercial PMs vary in speed and linearity of response, in the time fluctuations of the signal, in amplification factor (called *gain*), in the wavelength spectrum accepted, etc. A good intodurtory discussion can be found in [Leo94], and in the manufacturers' catalogues.

Pileup. Background signals which add to observed events, originating in multiple events that occur in the same time gate as signal of interest. Pileup is frequently seen in events observed at high luminosity colliders, where multiple collisions can even happen during a single bunch crossing of the collider.

Pitch. Pitch is typically taken to be the *geometrical pitch* of wires or strips in a planar detector, defined as the orthogonal distance between the centre of two adjacent parallel detector elements (the repetition unit). The spatial resolution of such a (one-dimensional) detector is determined by the pitch p: $\sigma \approx p/\sqrt{12}$;

The readout of the signal in a microstrip detector is often done by grouping several strips; the distance between two readout channels is then called the *readout pitch.* The readout pitch p_{r} is an integer multiple of the geometrical pitch p; the spatial resolution of a microstrip detector would then be $\sigma \approx p_{\mathrm{r}}/\sqrt{12}$; readout by charge division (→), however, can improve substantially the achievable accuracy (→ [Bates93], [Dabrowski96]).

Pixel Detector. A semiconductor detector made of wafers with very small rectangular two-dimensional detector elements, similar to pads (→), but of typical linear size of less than a mm. A large number of such detector elements on a surface ensures high spatial resolution in two coordinates, in the plane of the wafer. The precision achieved makes pixel detectors ideal candidates for vertex chambers, e.g. in experiments aimed at the detection of heavy-flavour particles, → [Hallewell96]. The readout of the very large number of channels is not without problems; it is usually done capacitively; → [Heijne89].

Planarity. → Jet Variables

PM. Short for Photomultiplier (→).

Proportional Counter, Proportional Mode. → Gaseous Detectors, Operational Modes

Proportional Tube. A drift tube (→) read out without measurement of drift time.

Pseudorapidity. The pseudorapidity is a handy variable to approximate the rapidity (→) if the mass and momentum of a particle are not known. It is an angular variable defined by

$$\eta = -\log\ \tan(\theta/2)$$

whose inverse function is

$$\theta = 2\tan^{-1}(\mathrm{e}^{-\eta}),$$

where θ is the angle between the particle being considered and the undeflected beam. η is the same as the rapidity y if one sets $\beta = 1$ (or $m = 0$). Statistical distributions plotted in pseudorapidity rather than rapidity undergo transformations that have to be estimated by using a kinematic model for the interaction.

Here is a table for the relation between θ and η for some round values:

$\theta[^0]$	90	45	40.4	15.4	15	10	5.7	2.1
η	0	0.88	1	2	2.03	2.44	3	4

Punchthrough. The longitudinal leakage of energy in a hadronic calorimeter. If the calorimeter is sufficiently deep to contain most showers, punchthrough is mostly due to non-interacting forward muons and/or neutrinos. Fluctuations in punchthrough result in low-end tails of measured energy distributions, and affect the resolution (→ e.g. [Fesefeldt90a]).

Quadrupole Magnet. Quadrupole magnets are characterized by a field of linear gradient:

$$B_z = Ky, \quad B_y = Kz$$

for two orthogonal coordinates y, z. Such a field results in equations of motion for a particle traversing the quadrupole:

$$\mathrm{d}^2y/\mathrm{d}s^2 = -ky, \qquad \mathrm{d}^2z/\mathrm{d}s^2 = kz$$

with k = normalized gradient = Kec/p (e = charge, c = speed of light, p = momentum).

This gives the *transport* equations:

$$\begin{aligned} y &= y_0 \cos a + 1/\sqrt{k}(dy_0/ds) \sin a \\ \mathrm{d}y/\mathrm{d}s &= -\sqrt{k}\ y_0 \sin a(\mathrm{d}y_0/\mathrm{d}s) \cos a \\ z &= z_0 \cosh a + 1/\sqrt{k}\ (dz_0/ds)\ \sinh a \\ \mathrm{d}z/\mathrm{d}s &= \sqrt{k}\ y_0 \sinh a + (\mathrm{d}y_0/\mathrm{d}s)\ \cosh a \end{aligned}$$

with $a = d\sqrt{k}$, d = length of quadrupole.

The focusing direction is y, with an approximate focusing length $f = 1/(kd)$, whereas in the z direction particles are defocused.

Racetrack. → Field Shaping

Radiation Damage in Gaseous Detectors. The main effect of prolonged operation of wire or microstrip gas chambers with many charged particles passing in the sensitive area, are deposits on the anode (e.g. sense wires), occasionaly also on the cathodes. Other effects concern walls and edges, and alterations in the gas composition, e.g. by *outgasing* from the chamber materials into the active gas. These *ageing effects* are highly dependent on the chamber geometry and on the operating conditions (mostly gas mixture and high voltage); they manifest themselves as films or protrusions (sometimes called "whiskers"), occasionally also as liquid depositions. The practical effect is loss of amplification ("gain") and the stretching of signals in time or other modifications of the signal shape; eventually, complete electrical breakdown (short circuit) is possible. Being due to polymerization, curing of damaged chambers a posteriori is difficult or impossible. Permanent gas circulation, as sometimes done in wire chambers, largely prevents radiation damage.

Microstrip gas chambers are affected by ageing in a similar way to wire chambers: outgasing and polymerization have the same effect.

A fairly complete review of wire chamber ageing can be found in [Kadyk91]; → also [Va'vra92]. For the ageing of MSGCs, → [Geijsberts92].

Radiation Damage in Plastic Scintillators. Radiation damage is the general alteration of the operational and detection properties of a detector, due to high doses of irradiation. There are three main aspects of radiation stability in plastic scintillators used as detectors: polymer hardness (optical stability), dopant stability, and stability of the fibre waveguide structure. The major role in the scintillation light losses are due to the bulk effects in the polymer. Considerable permanent absorption remains in the polymer as a result of irradiation. Several excited species can also be produced. The light yield is reduced by about 20% for a dose of 10^5 Gy. The transmittance loss is only about 2–6% at the same level (→ [Bross91]). Damage to dopants is of much less importance.

In scintillating fibres, radiation produces degradation of the scintillating core, polymer cladding and core/cladding interface effects. For reference, → also [Marini85].

Radiation Damage in Semiconductors. Radiation damage is the general alteration of the operational and detection properties of a detector, due to high doses of irradiation. In semiconductor devices, high-energy particles produce three main types of effects [Lint87]:

- Displacements. These are dislocations of atoms from their normal sites in the lattice, producing less ordered structures, with long term effects on semiconductor properties.
- Transient ionization. This effect produces electron–hole pairs; particle detection with semiconductors is based on this effect.
- Long term ionization. In insulators, the material does not return to its initial state, if the electrons and holes produced are fixed, and charged regions are induced.

Displacements determine the degradation of the bulk, and long term ionization is responsible for surface damage.

Producing displacements is a four-step process:

- The primary particle hits an atom in the lattice, transferring enough energy to displace it. Thus, interstitials and vacancies appear, and their pairing – the so-called Frenkel defects. In the case of high energies, nuclear reactions can occur, producing several fragments or secondary particles.
- The fragments of the target atom migrate through the lattice causing further displacements. The mean free path between two succesive collisions decreases towards the end of the range, so that defects produced are close enough and can interact.
- Thermally activated motion causes rearrangement of the lattice defects at room temperature (annealing). Part of these rearrangements are influenced by the presence of impurities in the initial material.
- Thermally stable defects influence the semiconductor properties, i.e. also the detector parameters.

Effects of displacements are to be seen in the increase of capture, generation and recombination rates of the non-equilibrium charge carriers. In detectors they cause changes of the internal electric field, due to the modified doping concentration, going eventually up to inverting the conduction type for very high irradiations, increase of the leakage current, changes in capacitance and resistivity, and charge collection losses.

Long term ionization effects also comprise several steps:

- Ionization is produced along the track of the primary ionizing particle, or sometimes in restricted regions around a nuclear reaction. Electrons and holes are created, with a certain distribution.
- Many of the e-h pairs produced recombine before they could move due to diffusion or the electric drift. Recombinations take place between particles produced in the same or in different events.
- The electrons which did not recombine in the initial phase diffuse or drift away. Some electrons end up on traps, others may escape from the insulator.
- The carriers trapped on levels with low ionization energies are thermally reexcited and get into the conduction or valence band; they are subject to further drift or diffusion, and leave the dielectric or are captured on deep trap levels (practically permanent).
- Apart from the production of trapped charge, in the energy gap new oxide–silicon interface levels are induced. These interface states are occupied by electrons or holes, depending on the position of the Fermi level at the interface.
- The net effect of the induced charges in the oxide is the change of the electric field in the semiconductor, in the vicinity of the interface.

All effects depend on the particle type and the incident energy.

Radiation Length. Scaling variable for the probability of occurrence of bremsstrahlung (→) or pair production (→), and for the variance of the angle of multiple scattering (→). The radiation length is given by

$$1/X_0 = \frac{4\alpha N_{\mathrm{A}} Z(Z+1)\, r_{\mathrm{e}}^2 \log(183\ Z^{-1/3})}{A}$$

with

$$\alpha = \text{fine structure constant} (\approx 1/137)$$

$$N_{\mathrm{A}} = \text{Avogadro's number } (6.022 \cdot 10^{23}/\text{mole})$$
$$Z = \text{atomic number of the traversed material}$$
$$A = \text{atomic weight of the traversed material}$$
$$r_{\mathrm{e}} = \text{electron radius } (2.818 \cdot 10^{-13}\ \text{cm}).$$

A detailed discussion can be found in [Rossi65] or in [Jackson75], → also [Kleinknecht82]; for numerical values, → [Barnett96].

The average energy loss due to bremsstrahlung (→) for an electron of energy E is related to the radiation length:

$$-(\mathrm{d}E/\mathrm{d}x)_{\mathrm{brems}} \hat{=} E/X_0 \ ,$$

and the probability for an electron pair to be created by a high-energy photon is $7/9\ X_0$.

A brief discussion, more references, and a table of radiation lengths for various materials are also given in [Barnett96].

Radiation Measures and Units. The term *dose* (correct form *absorbed dose*) D refers to the mean energy imparted[2] by ionizing radiation to the matter in a volume divided by the mass contained in the respective volume [ICRUReport33]:

$$D = \frac{\mathrm{d}\bar{\varepsilon}}{\mathrm{d}m}$$

where $\mathrm{d}\bar{\varepsilon}$ is the mean energy imparted by ionizing radiation to matter of mass $\mathrm{d}m$.

The *KERMA* K (also *kerma*, short for *kinetic energy released in matter*) is the quotient of $\mathrm{d}E_{\mathrm{tr}}$ by $\mathrm{d}m$, where $\mathrm{d}E_{\mathrm{tr}}$ is the sum of the initial kinetic energies of all the charged ionizing particles liberated by uncharged ionizing particles (indirectly ionizing particles) in a material of mass $\mathrm{d}m$ [ICRUReport33]:

$$K = \frac{\mathrm{d}E_{\mathrm{tr}}}{\mathrm{d}m}$$

---

[2] The energy imparted ε is equal to the difference between the radiant energy incident on and emerging from the volume (excluding rest mass energies) of all the charged and uncharged ionizing particles entering the volume, plus the sum of all changes (decrease: positive sign, increase: negative sign) of the rest mass energy of nuclei and elementary particles in any nuclear transformations which occur in the volume. ε is a stochastic quantity. Its expectation value $\bar{\varepsilon}$ is termed the *mean energy imparted*, and it is non-stochastic.

The KERMA characterizes the interaction of uncharged particles.

Both absorbed dose and KERMA are measured (in SI, → Units) in $\mathrm{J\,kg^{-1}}$; the special name for this unit is the gray (Gy). An older unit, still in restricted use, is the rad: 1 rad = 10^{-2} Gy. → also NIEL Scaling.

The *(particle) flux*, $\dot{N}$, is the quotient of $\mathrm{d}N$ by $\mathrm{d}t$, where $\mathrm{d}N$ is the increment of particle number in the time interval $\mathrm{d}t$.

$$\dot{N} = \frac{\mathrm{d}N}{\mathrm{d}t} .$$

The particle flux is measured in $\mathrm{s^{-1}}$.

The *(particle) fluence*, Φ, is the quotient of $\mathrm{d}N$ by $\mathrm{d}a$, where $\mathrm{d}N$ is the number of particles incident on a sphere of cross-sectional area $\mathrm{d}a$

$$\Phi = \frac{\mathrm{d}N}{\mathrm{d}a} .$$

The SI unit for the fluence is $\mathrm{m^{-2}}$, but the frequently used unit is $\mathrm{cm^{-2}}$.

The *(particle) fluence rate*, φ, is the quotient of $\mathrm{d}\Phi$ by $\mathrm{d}t$, where $\mathrm{d}\Phi$ is the increment of particle fluence in the time interval $\mathrm{d}t$:

$$\varphi = \frac{\mathrm{d}\Phi}{\mathrm{d}t} = \frac{\mathrm{d}^2N}{\mathrm{d}a\mathrm{d}t} .$$

The fluence rate is measured in $\mathrm{m^{-2}s^{-1}}$.

The term particle flux density is also in use for the same quantity; the preferred name is particle fluence rate.

The *activity*, A, of an amount of radioactive nuclide in a particular energy state[3] at a given time is the quotient of $\mathrm{d}N$ by $\mathrm{d}t$, where $\mathrm{d}N$ is the expectation value of the number of spontaneous nuclear transitions from that energy state in the time interval $\mathrm{d}t$:

$$A = \frac{\mathrm{d}N}{\mathrm{d}t}$$

The activity is measured in $\mathrm{s^{-1}}$, the unit having the special name becquerel (Bq): $1\,\mathrm{Bq} = 1\,\mathrm{s^{-1}}$. A special unit also still in use is the curie (Ci): $1\,\mathrm{Ci} = 3.7 \times 10^{10}\ \mathrm{s^{-1}}$ (exactly).

The activity of an amount of a radioactive nuclide in a particular energy state is equal to the product of the decay constant for that state and the number of nuclei in that state.

[3] in the ground state unless otherwise specified

A detailed discussion and definitions on radiation quantities and units can be found in [ICRUReport33].

Radiative Corrections. The corrections to theoretical predictions in high-energy physics which are due to Feynman diagrams containing additional emitted field quanta, usually photons or gluons. The additional diagrams may contain loops, i.e. reabsorption of virtual quanta (vertex corrections, vacuum polarization), or describe bremsstrahlung-like emission of real quanta (internal bremsstrahlung).

The radiative corrections are strongly dependent on the reaction studied. For literature and an available program for e^+e^-, → [Berends83] and [Berends82]. Expressions for ep and νp scattering are found in [Maximon69], [Mo69] or [Bjorken63]. For more recent developments, → [Jadach97] and [Ward95].

Range. The range of a particle of given energy in an absorbing material is the average thickness of material (usually defined parallel to its original direction) which it traverses before coming to rest. For a charged particle other than an electron, this is dominated by energy loss (→) due to ionization; for electrons, the energy loss due to bremsstrahlung (→) is more relevant. Curves and tabulations of the range due to ionization loss exist for different particles in different absorbers (→ [Barnett96], [Serre67], [Richard71]). Fluctuations in the energy loss result in a distribution of range values for particles with the same energy, which is called range straggling (→).

Rapidity. The rapidity is a variable frequently used to describe the behaviour of particles in inclusively measured reactions. It is defined by

$$y = \frac{1}{2} \log \frac{E + p_{\mathrm{L}}}{E - p_{\mathrm{L}}}$$

which corresponds to

$$\tanh(y) = p_{\mathrm{L}}/E \ .$$

y is the rapidity, p_{L} is the longitudinal momentum along the direction of the incident particle, E is the energy, both defined for a given particle. The accessible range of rapidities for a given interaction is determined by the available centre-of-mass energy and all participating particles' rest masses. One usually gives the limit for the incident particle, elastically scattered at zero angle:

$$|y|_{\max} = \log[(E+p)/m] = \log(\gamma + \gamma\beta)$$

with all variables referring to the through-going particle given in the desired frame of reference (e.g. in the centre of mass).

Note that $\partial y/\partial p_{\mathrm{L}} = 1/E$. A Lorentz boost β along the direction of the incident particle adds a constant, $\log(\gamma + \gamma\beta)$, to the rapidity. Rapidity differences, therefore, are invariant to a Lorentz boost. Statistical particle distributions are flat in y for many physics production models.

Frequently, the simpler variable *pseudorapidity* (→) is used instead of rapidity (and sloppy language mixes up the two variables).

Rayleigh Scattering. → Compton Scattering

Relativistic Rise. From $\beta\gamma = p/mc \approx 4$ upwards, the energy loss (→) of particles traversing a medium starts to increase logarithmically because of relativistic effects (up to $\beta\gamma \approx 3$, the loss of energy decreases with $1/\beta^2$ as expected even from a classical point of view where energy loss is related to the number of collisions of the traversing particle with the atoms of the medium). At high values of $\beta\gamma$ ($>$ 100), the rate of ionization energy loss saturates due to density effects (*Fermi plateau*). In noble gases with high Z, the Fermi plateau is about 1.5 to 1.7 times the minimum ionization (→).

With knowledge of the particle momentum, the relativistic rise can be used for hadron identification over a wide range of momenta; taking π and K mesons and protons as an example: π/K separation using the relativistic rise is possible from 2 to 50 GeV/c, K/p from 5 to 40 GeV/c. In argon (at 0° temperature and 760 mm pressure Hg) the relative ionization for particles of 10 GeV/c is

$$I_p : I_k : I_\pi = 109 : 118 : 136 \quad (100 = \text{minimum ionization})\ .$$

For details, → [Allison91]. For an example of measuring the relativistic rise, → [Breuker87].

Resolution. → Spatial Resolution or Time Resolution

RICH. Short for *Ring Imaging Cherenkov Counter* (→)

Ring Imaging Cherenkov Counter. A large-acceptance detector using photons from Cherenkov radiation for a measurement of the particle velocity β. Particles pass through a radiator, the radiated photons may be directly collected by (or are focused by a mirror onto) a position-sensitive photon detector. Respectively, these are called *direct focusing* or *mirror-focused* RICH detectors. For direct focusing, radiators have to be kept thin (e.g. a liquid radiator), to avoid broadening the ring or filling it; however, [Fabjan95b] report a use of a similar setup as a threshold counter. The Cherenkov radiation emitted at angle δ is focused onto a ring of radius r at the detector surface, and β can be determined by a measurement of r. For photon detection one uses thin photosensitive (an admixture of e.g. triethylamine to the detector gas) proportional or drift chambers, → [Barrelet91].

A detailed treatment of errors in Cherenkov detectors can be found in [Ypsilantis94]. An outlook for the future use is given in [Treille96].

For the various currently successful ways of building practical RICH detectors, → [Ekelof96] or [Ypsilantis94], and literature given there. An example is the combined RICH with liquid radiator (unfocused) and gas radiator (mirror-focused) of the DELPHI experiment at LEP (→ [Abreu96], [Aarnio91]):

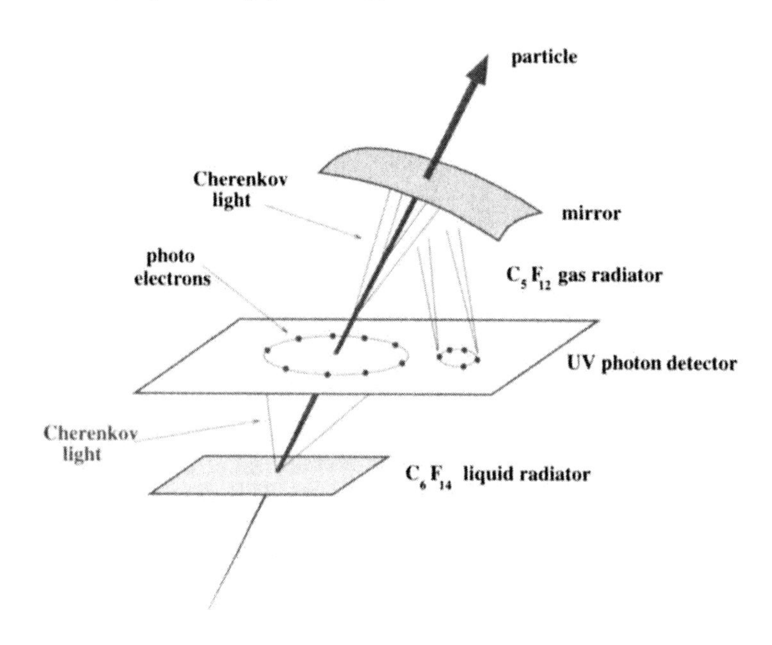

A combined tracking-cum-RICH project, including even identification of particles by energy loss, has been described in [O'Brien91]. For using a RICH as a significant trigger device, → [Baur94].

Scaling Variable. One calls a scaling variable any variable allowing one to find an identical description of phenomena in different situations by expressing them in a suitable projection. Examples are the description of electromagnetic interactions in matter in terms of L/X_0 where X_0 is the radiation length (→), the probability distribution of multiplicities of produced particles which *scales* approximately if plotted in terms of $z = n/\langle n \rangle$ (→ KNO-distribution), and in particular the scaling variables in inclusive reactions (→ Deep Inelastic Scattering Variables).

Deviations from exact scaling are called *scaling violations.*

Scintillating Fibre Calorimeter. A heterogeneous calorimeter whose cells are volumes filled with a reasonably homogeneous arrangement of scintillating fibres interspersed with some passive material. Fibre calorimeters are comparatively cheap, produce a fast signal, and give good energy resolution due to very frequent sampling. They have been used for electromagnetic calorimetry at moderate energy ([Hertzog90]). Advantages and limitations are discussed in [Livan95].

Scintillating Fibre Tracker. Detectors based on large numbers of fibres arranged in volumes, individually producing scintillation light at the passage of a charged particle. The fibres can be arranged in layers, and essentially act like arrangements of proportional tubes. The relatively modest light yield and cross-talk between adjacent fibres are problematic; also, the readout via photomultipliers is non-trivial. A possible readout via avalanche photodiodes has been discussed in [Nonaka96]. Resolutions of ±200–300 μm or better have been reported; → [Aschenauer97], [Agoritsas95].

Scintillation Counter. Scintillation counters have been in use since the beginning of the century, making use of the property of certain chemical compounds to emit short light pulses after excitation by the passage of charged particles or by photons of high energy. Scintillation is characterized by the light yield (→), the absorption and emission spectrum (→ Wavelength Shifter), signal linearity, and the pulse shape, viz. rise and decay times; the latter range from less than 1 ns (modern fast plastic scintillators) up to 5 μs (for RbCl). The numbers given in the literature vary considerably even for the same material, due to different surface treatments; ageing and radiation damage account for additional fluctuations.

Scintillating material can be *organic* (solid crystals, plastics, i.e. synthetic polymers, or liquids), or *inorganic* (crystals or glasses); also gaseous scintillators are in use. Examples of organic crystals are anthracene ($C_{14}H_{10}$), trans-stilbene ($C_{14}H_{12}$), or naphtalene ($C_{10}H_8$); organic liquids usually have brand names (PPO, POPOP, NE213, PBD, etc). Among plastics, two- or three-component scintillators are common, with a solid solvent, doped with aromatic compounds (TP, PPO or PBD) or with wavelength shifters; polysterene and polyvinyltuolene are most commonly used. Inorganic crystals include Na(Tl), CsI(Tl), BGO (→), and BaF_2; high-Z crystals make good high-energy physics scintillators, and are used in crystal calorimeters (→). As gaseous or liquid scintillators one uses Xe, Kr, Ar, He, or N.

In organic scintillators, ionizing particles provoke an excitation of molecular levels, which causes light in the UV region to be emitted. Added wavelength-shifting molecules absorb the UV photons and reemit visible light, in the blue region (around 400 nm wavelength). Inorganic materials are frequently doped with other materials acting as an activator centre by the capture of holes or electrons generated by ionization.

For historical reasons, anthracene ($C_{14}H_{10}$) is used as a standard for the light gain. The *absolute* scintillation efficiency of anthracene crystals is of the order of 0.05, and is discussed in [Brooks79] and [Birks64]. The most commonly used inorganic scintillator in nuclear physics is NaI(Tl) (NaI doped with Tl), because of its good energy resolution. As particle physics detector, NaI is not popular, being hygroscopic, difficult to machine, comparatively slow and expensive. All the same, inorganic scintillators are often compared to NaI(Tl), whose absolute scintillation efficiency is about 0.1 (or about 1 photon/25 eV, → [Heath79]).

For a good overview, → [Bicron93]). An introduction to (organic) scintillators can be found in [Zorn92]; [Doke91] discusses the scintillation of noble liquids.

In high-energy physics experiments, scintillation counters are used for timing (*time-of-flight counters*), or for fast event selection (*trigger counters*, or groups of counters connected by fast logic into *hodoscopes*); they are also vital for measuring the energy of particles by total absorption in sampling calorimeters (→), which is possible due to the proportionality of light output to the energy loss of the particle. Inorganic materials are popular for high-precision calorimetry (→ Crystal Calorimeter). In large scale calorimeters, wavelength shifters are also used as light collecting devices [Bourdinaud81].

Semiconductor Detectors. Semiconductor detectors have been used in high-energy physics applications in the form of pixel detectors (→), microstrip detectors (→) and pads (→); they are popular due to their unmatched energy and spatial resolution, and have excellent response time. These detectors are manufactured mainly of silicon, traditionally on high-resistivity single crystal float-zone material. GaAs is perhaps a future alternative to silicon; presently, it seems to be an expensive and not fully mastered technology of potentially better radiation hardness (→ [Smith96], [Chmill94], [Chmill93], [Beaumont90]).

Similar structures have been proposed on diamond, too, another possible candidate for detectors in the future (→ [Bauer96]).

After the first implementation of a planar technology in 1980 [Kemmer80], semiconductors were quickly understood to give detectors of extraordinarily high performance. Recent progress in microtechnology now allows reliable large-scale production of detectors of sophisticated designs, at acceptable cost; their properties have been pushed to:

- position localization accuracy of 5 μm in one coordinate,
- two-track separation down to 10 μm,
- geometrical accuracy in the region of 1 μm,
- bias voltages less than 100 V for microstrip detectors,
- time response less than 5 ns,
- relatively simple installation.

The principle of operation of a semiconductor detector is the following: if an ionizing particle penetrates the detector it produces electron–hole pairs along its track, the number being proportional to the energy loss. An externally applied electric field separates the pairs before they recombine; electrons drift towards the anode, holes to the cathode; the charge is collected by the electrodes (charge collection). The collected charge produces a current pulse on the electrode, whose integral equals the total charge generated by the incident particle, i.e. is a measure of the deposited energy. The readout goes through a charge-sensitive preamplifier, followed by a shaping amplifier. Silicon detectors are asymmetric p-n junctions; to work as a detector, the p^+n diode is reverse-biased by applying a positive voltage on the rear ohmic contact, a metal deposited on the n side. At full depletion, the electric field is a maximum in the junction and decreases to zero at the ohmic contact. In order to avoid losses in charge collection, the silicon detectors are overbiased (below break-down voltage).

The intrinsic energy resolution is related to the low energy threshold: only 3.6 eV are necessary to produce an electron–hole pair, a low value compared to the ionization energy in a gas (30 eV) or the approximately 300 eV necessary to extract an electron from a photocathode coupled to a plastic scintillator. The good spatial resolution comes from the high density of Si, which reduces the range of the secondary electrons. On the other hand, the average energy loss in Si is high, about 390 eV/μm [Barnett96], for a <1 1 1> oriented single crystal, and corresponds to about 110 e-h pairs. To limit the multiple Coulomb scattering, the detector thickness must be kept thin; the usual compromise thickness is $\sim$ 300 μm for optimum detection. The thickness of the detector also determines the amplitude of the signal, as there is no charge multiplication in silicon; the signal–to–noise ratio is, therefore, a critical issue. For 300 μm, one gets on average 3.2×10^4 electron–hole pairs, a signal requiring low-noise electronics.

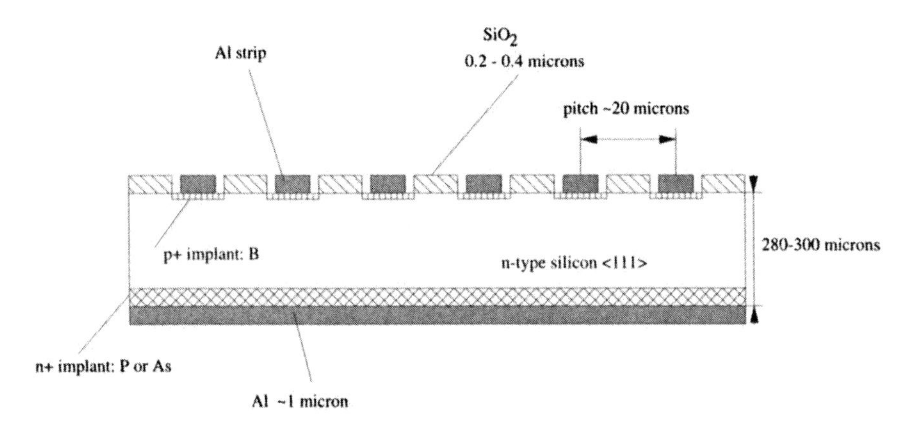

The high particle fluence in the interaction regions of colliders is a critical parameter for the operational parameters and detector performance (→ Radiation Damage in Semiconductors).

The crystals (wafers) from which the detectors are composed, are produced by specialized companies. Silicon is a IV group element in the periodic table. In the intrinsic material the electron and hole densities are equal; at room temperature $n_i = 1.45 \times 10^{10}$ cm^{-3}. Materials of p or n type are obtained by replacing some silicon atoms with atoms of the III group, or V group, respectively.

Doping, via the charge carrier concentration, determines the resistivity ρ of the semiconductor material. Detector-grade silicon has very low doping, i.e. high resistivity ($N_{\text{eff}} \leq 10^{12}$ cm^{-3}, $\rho \geq 2$ kΩ cm). Float-zone n-type material is produced from silicon with minimum boron concentration, by repeated zone refining to reduce the phosphorus concentration [Dreier90].

Nuclear physicists widely use P–I–N structures along with p$^+$n diodes. Doped detectors like Ge(Li), Si(Li) are also common. Surface barrier detectors (e.g. Au deposited on silicon) have been used in calorimetry, e.g. at H1 at HERA, → [Fretwurst96].

Basic references to semiconductor devices are [Lutz95] and [Sze81].

Sense Wire. An anode wire in multiwire chambers (→) on which the avalanche of electrons is collected (→ Gaseous Detectors, Operational Modes). In drift chambers (→), sense wires are usually sandwiched between two field wires. Sense wires are generally thin (of the order of 20–30 μm diameter) to have a high gradient in the electric field very close to the wire (multiplication region), so that an electron avalanche is produced at not too high a value of the voltage on the cathodes. Sense wires are sometimes resistive wires to measure charge division.

Shimming of Magnets. The addition of small pieces of magnetizeable iron (a *shim* is a piece of corrective material) in order to achieve desirable properties of a magnetic field, e.g. to approximate better

- constant deflection power $\int B \, dl$ (→ Errors in Track Reconstruction),
- field smoothness on the edges of pole faces,
- small field gradients in the neighbourhood of a symmetry plane,
- cylindrical symmetry for cylindrical pole faces in a C-shaped yoke,
- correction of imprecisions of coil mounting.

Shower. Cascade of secondary particles produced in interactions of high-energy particles in dense matter. (→ Calorimeter, Electromagnetic Shower, Hadronic Shower).

Silicon Drift Chamber. A detector based on wafers of semiconductors and aiming to obtain the two-dimensional capabilities and the resolution of the pixel detector (→), but circumventing the problem of the large number of channels. This detector essentially consists of a fully depleted thin semiconductor wafer, with a linear arrangement of anodes at the edge (or on two edges opposite each other), and a cathode at the opposite edge (or in the middle). An electric field makes the electrons, generated by a passing charged particle, drift towards the array of anodes, at a moderate speed (like 15 mm/μs), thus obtaining two coordinates in the plane of the wafer: the anode position along the edge, and the drift time converted to distance perpendicular to the edge. The drift length is limited by the size of the wafer (a few centimetres), the anode pitch is chosen to be submillimetre, and a resolution of ± 100 μm can be achieved (charge division between anodes can improve the precision). Early references are [Gatti88] and [Gatti84], more recent ones are [Vacchi93] or the specialized meeting proceedings [Holl96].

Simulation of Showers. Although the basic physical processes occurring in electromagnetic showers are well known, this is not quite so for hadronic showers. The simulation of showers in calorimeters needs to follow all particles to rather small energies; for hadrons, phenomenological approximations for intra-nuclear cascades and intermediate-energy processes have to be made (→ [Ferrari93], [Ferrari97]), and also electromagnetic simulation results can be sensitive to multiple low-energy cutoff parameters. The number of particles in a shower is very large, particularly at high energies, so that even the computing resources of large laboratories can be challenged by full simulation programs. Simulation being a central tool in optimizing calorimeters, the validity of results and questions of efficiency on computers has led to multiple publications and comparisons (e.g. [Fesefeldt90b]). Much overall understanding can be derived from *average shower parameters* (→ Electromagnetic Shower, Hadronic Shower). *Full shower simulation* programs are EGS (→ [Nelson85]) for electromagnetic showers, or GEISHA (→ [Fesefeldt85], [Fesefeldt90a]) or FLUKA (→ [Fassò93]) for both electromagnetic and hadronic showers. One should note that multiple versions for these programs exist, and that they are usually applied embedded in programs allowing one to introduce the geometry of a detector. Recently, many initiatives exist to allow the detector designers' form of describing detector components (in a CAD program's data base) to enter directly the simulation programs (which simulate not only showers, but also tracking detectors). Up-to-date manuals and specialist expertise are needed for more information.

The tuning of the simulation parameters, in particular of cutoffs and integration step sizes, is delicate and depends on the goal of the simulation; some codes (e.g. FLUKA) use mathematical methods (like importance sampling) to achieve robustness and speed up calculation, alleviating somewhat this tuning problem. In major simulation projects at high energies users have also resorted to relatively high cutoff parameters, using for further shower development randomly selected showers at lower energies, precomputed in full simulation and stored in "libraries" (→ [Graf90]). This strategy can save substantial amounts of computer time.

Single Electron Peak. Distribution of the total charge observed at the anode of a photomultiplier, resulting from a single electron leaving the cathode. For the single electron peak the spread in transit time can be assumed to be independent of the gain fluctuations.

The single electron peak is useful in studying the mean gain and the statistical properties of a photomultiplier (→). The gain is given by the abscissa of the centre of gravity in units of the electron charge, and the relative variance can be approximated by

$$v_{\mathrm{sep}} = v_{d1} + v_{d2}/g_1 + v_{d3}/g_1 g_2 + \dots ,$$

where v_{di} is the relative variance of the gain at the ith dynode, and g_i is the average gain at the ith dynode (100% interstage collection efficiency is assumed).

In practice, the single electron peak is measured by illuminating the photocathode by a triggered light source attenuated by suitable filters, and electrons leaving the cathode are Poisson distributed with an average

$$N \equiv \langle n_{\mathrm{pc}} \rangle = \langle n_{\mathrm{lightsource}} \rangle \cdot \eta_{\mathrm{pc}}$$

where η_{pc} is the averaged conversion efficiency of the photocathode.

The contribution of zero electrons, therefore, is

$$p(0) = \mathrm{e}^{-N} ,$$

contributing only the dark current (→) integrated over the gate length of the analogue-to-digital Converter.

For $n_{\mathrm{pc}} \ll 1$ the contribution of more than one electron is given by

$$p(n > 1) \approx \langle \bar{n}_{\mathrm{pc}} \rangle^2 / 2 \, .$$

At low flux ($\lambda \ll 1$) one finds

$$\begin{aligned} P(0e^-) &= \mathrm{e}^{-\lambda} && \approx 1 - \lambda \\ P(1e^-) &= \lambda \mathrm{e}^{-1} && \approx \lambda \\ P(2e^-) &= \lambda^2 \mathrm{e}^{-\lambda}/2 && \approx \lambda^2/2 \, . \end{aligned}$$

To evaluate the gain of the photomultiplier one has to correct for the collection efficiency η of the first dynode. To evaluate the photon flux reaching the cathode the quantum efficiency of the photocathode must also be considered (up to 25% and dependent on the wavelength).

Space Charge. In a proportional or drift chamber, positive ions are released during the amplification process. While the negative charges (electrons) are collected on the anode wire, the ions drift slowly towards the cathode. In normal operational conditions (→ Gaseous Detectors, Operational Modes), the charge density in the electron–ion avalanche is small compared to the charge density on the wire, and only small signal distortions occur. For chambers exposed to high-intensity beams, they may accumulate, and then produce space charge effects, in particular distortions of the effective electric field, causing inefficiencies of the chamber and thus influencing acceptance and calibration parameters of the chamber (e.g. causing a non-linear relation between the collected charge and $\mathrm{d}E/\mathrm{d}x$. For a mathematical treatment, → [Sipilä80]. For more on the charge buildup in wire chambers, → [Sauli91], [Blum93].

Compare the space charge in wire chambers with the space charge region in a semiconductor detector, which is, in fact, the active region (the depleted volume).

Spark Chamber. A historic device using electric discharges over a gap between two electrodes with large potential difference, to render passing particles visible: sparks occurred where the gas had been ionized. Most often, multiple short gaps were used, but wide-gap chambers with gaps up to 40 cm were also built. Particle trajectories had to follow the electric field to within $\pm 40^o$. For detailed reading, → [Shutt67].

Mostly, the sparks were photographed and analysed from film, but also acoustic chambers with recordings via transducers were built. Eventually, the drive towards digitized information resulted in electrodes built as wire grids, and the additional change in operational mode resulted in modern gaseous chambers.

Spatial Resolution. In particle tracking detectors, "resolution" is most commonly used in a double sense:

a) the precision of localizing a track, or, more precisely, the square root of the variance of the conditional probability density function to get a signal x when a particle has crossed the detector at x':

$$\sigma(x') = \left[\int (x - x')^2 r(x; x')\,\mathrm{d}x\right]^{1/2} ;$$

b) the capability of a detector to separate the signals arising from two particles traversing the detector at close distance.

In practice, the function $r(x; x')$ can often be approximated by another function:

$$r(x; x') \approx r'(x - x') .$$

In this case σ no longer depends on (or depends only very smoothly on) x', and can be considered to be constant during the fitting of an individual track. An example where this does not hold is a dense stack of proportional chambers (→ [James83]).

In rigorous track fitting, the spatial resolution must be considered in conjunction with the effect of multiple scattering (→).

Spatial resolution, or at least position resolution in the transverse direction, is also a property of calorimeters (→); in particular, electromagnetic showers need to be localized in order to associate track information from other detectors. Also, for jets, calorimetric information is often used alone to determine effective masses; here, again, positional information has to be available.

Sphericity and Spherocity. → Jet Variables

Stokes Shift. The difference in wavelength between absorbed and emitted quanta, in wavelength shifters or scintillators. The emitted wavelength is always longer (if single photons are absorbed) or equal to the incident wavelength, due to energy conservation; the difference is absorbed as heat in the atomic lattice of the material.

Straggling. The range (→) of a charged particle in matter is subject to fluctuations caused by the probability distribution of energy loss (→). These fluctuations are called range straggling; compare this to energy loss straggling, which describes the fluctuations in energy loss. Range straggling can be described by

$$\sigma_{\mathrm{R}}/R = f(\gamma - 1)/\sqrt{(M)}$$

where R is the range, $\gamma = 1/\sqrt{(1-\beta^2)} = E/Mc^2$, E and M are the particle energy and mass, and f is a slowly varying function depending on the absorber (→ [Rossi65]). For protons in liquid hydrogen, the relative error σ_{R}/R is of the order of 0.015.

Straw Chamber. → Drift Tube

Streamer. → Gaseous Detectors, Operational Modes

Streamer Chamber. An optical chamber using the fact that streamer discharges can be controlled by pulsing the field; such discharges occur along the path of an ionizing particle, in a gas subjected to an electric field (→ Gaseous Detectors, Operational Modes). The resulting beginnings of sparks can be recorded optically, and the recorded information analysed off-line. Chambers with a sensitive volume of several cubic metres were built, using fields of extremely short duration ($\approx$ 15 ns FWHM).

The technique competed with that of bubble chambers, producing images of excellent quality over large volumes, and had the advantage of being triggerable by external devices (via the pulsed electric field). Streamer chambers were used in colliding beams (→ [Rushbrooke81]) and in fixed-target experiments (e.g. [Teitelbaum92]), the collision point or target being outside the chamber. The need for optical recording and the associated dead time eventually turned out to favour electronic techniques like drift chambers (→).

Streamer Tube. → Limited Streamer Tube

Structure Function. In a simplified view, structure functions allow one to derive the distribution of partons inside a composite particle (hadron), i.e. the probability density function of the fraction of momentum carried by constituents like quarks or gluons. Structure functions have been a most useful concept in deep inelastic scattering of leptons on nucleons, where the constituent hypothesis of hadrons was first verified (→ Deep Inelastic Scattering Variables). The differential cross-section $\mathrm{d}^2\sigma/\mathrm{d}Q^2\mathrm{d}x$ of deep inelastic scattering is, in fact, expressed in terms of structure functions. A model is needed to relate the structure functions to momentum distributions (the quark-parton-model, → [Close79]); measurements reveal that the structure functions not only depend on the quark flavour (u and d in protons and neutrons), but also on the probing particle. For a review on the experimental situation, → [Martin93], [Martin94], [Abramowicz96].

Structure functions (of the photon) are also used in *two-photon interactions*, a type of e^+e^- interaction which can be viewed as electron–photon scattering.

Synchrotron Radiation. Radiation emitted by an electric charge travelling in a magnetic field, due to transverse acceleration. The total energy loss is given by

$$\mathrm{d}W/\mathrm{d}t = (2c/3)e^2\beta^4[E/(m_0c^2)]^4(1/r^2)\ ,$$

with

$$\begin{aligned} m_0 &= \text{rest mass of particle} \\ E &= \text{energy of particle} \\ r &= \text{bending radius.} \end{aligned}$$

Due to the factor $[E/(m_0c^2)]^4 = \gamma^4$, synchrotron radiation is mainly observed with low-mass particles, e.g. as a major energy loss in electron ring accelerators.

The mean energy loss of electrons in a circular orbit due to synchrotron radiation is (per revolution)

$$W_\mathrm{r} = t_\mathrm{r}(\mathrm{d}W/\mathrm{d}t) = (4\pi/3)e^2\beta^3\gamma^4/r = CE^4/r$$

with

$$\begin{aligned} t_\mathrm{r} &= \text{revolution time} = 2\pi r/(c\beta)\ , \\ C &= 8.85 \times 10^{-5} \quad [\mathrm{GeV}^{-3}/\mathrm{m}]\ . \end{aligned}$$

In a machine with bending and straight sections, the energy loss per revolution is independent of the straight sections. In LEP, with a bending radius $r = 3100$ m and bending sections over $L_\mathrm{b} = 19\,500$ m, the radiated energy per revolution and particle at collision energy

$\sqrt{s} = 92$ GeV is $W_r = 180$ MeV, the energy per second and particle ($t_r = 89$ μs) is $dW/dt = 2000$ GeV/s.

In electron colliders, synchrotron radiation limits the bending power that can be installed and imposes a lower limit on the ring radius; the effect has, on the other hand, been used to determine and monitor beam parameters in accelerators. On CERN's projected Large Hadron Collider (proton–proton), synchrotron radiation will cause up to 4 kW loss per beam.

Photons radiated as synchrotron radiation have a broad energy spectrum at low energies; above a *critical energy* $\epsilon_c = 3\hbar c\gamma^3/2r$ the falloff is exponential. ϵ_c is also the median of the power distribution, viz. an equal amount of energy is radiated below and above ϵ_c. Radiated photons are concentrated in a forward cone with opening angle (FWHM)

$$\Theta \approx \frac{2}{\gamma}\sqrt{\frac{\epsilon_c}{\epsilon}}$$

(for $\epsilon \ll \epsilon_c$, replace $\sqrt{\ }$ by $\sqrt[3]{\ }$).

Synchrotron light sources are used in many specialized laboratories around the globe for a variety of applications, like in medicine, for materials research, structural molecular biology, etc.

Thomson Scattering. → Compton Scattering

Thrust. → Jet Variables

Time Expansion Chamber. First proposed by Walenta ([Walenta79b], [Walenta82]), the time expansion chamber is a special type of time projection chamber (→ Drift Chamber), with its drift volume arranged to achieve higher precision: the drift volume is separated into two distinct parts, a drift region of relatively low electric field, with consequently reduced drift velocity, and a higher-field amplification region close to the wire, which is necessary to obtain a high enough signal. In practice, this separation is achieved by inserting a grid of defined potential between the drift region and the sense wires; the sketch of principle (→ Drift Chamber) is the same for a time expansion chamber, but the function of the grid is different.

Drift velocities in the time expansion chamber are reduced typically by one order of magnitude, and resolutions of ±40 μm have been reported ([Anderhub88]).

Time Projection Chamber. A drift chamber with large drift volume, and drift direction perpendicular to the plane of the sense wires. → Drift Chamber.

Time Resolution. Time resolution is the criterion for the quality of a time measurement, e.g. in a fast trigger scintillator. In general, the standard deviation is used to describe the resolution, i.e. the root of the variance. It plays a role in fast triggering, particularly for time-of-flight measurements. The time resolution can be influenced by the material, the size and the surface of the scintillation counter (→), the properties of the light guide, the fluctuation in gain and the spread in transit time of the photomultiplier, and the threshold of the discriminator. In the case of large scintillation counters, a "meantimer" circuit averaging over the arriving time of the light signals at either end of the scintillator can correct for the delay due to the distance the light has to travel in the scintillator. Typically the time resolution in time of flight measurements varies from 0.1 to 1 ns for high-quality scintillators.

In drift chambers (→), spatial resolution (→) and time resolution are closely connected; the term resolution there is used in a double sense: it describes the effective resolution achieved on measuring points on an isolated track, and the resolution of two points from two nearby tracks.

TOF Counter. Short for time-of-flight counter, → Scintillation Counter

TPC. Short for time projection chamber, → Drift Chamber

Trajectory of a Charged Particle. The trajectory of a charged particle can be calculated from the equations of motion. Neglecting radiative corrections (→ [Jackson75]) and in the absence of an electric field, the equation has the simple geometrical form:

$$\mathrm{d}^2\boldsymbol{x}/\mathrm{d}s^2 = (q/|\boldsymbol{p}|)[(\mathrm{d}\boldsymbol{x}/\mathrm{d}s) \times B(\boldsymbol{x})]$$

For details and units → Equations of Motion.

The integral of this second-order differential equation depends on six initial values. Assuming an s_0 on a reference surface (e.g. a plane z = const.), the trajectory is determined by five parameters, e.g. x, y, $\mathrm{d}x/\mathrm{d}s$, $\mathrm{d}y/\mathrm{d}s$ and $1/|\boldsymbol{p}|$ in the reference plane.

For constant $\boldsymbol{B}$ the solution of the equations is a helix. Choosing $\boldsymbol{B} = (0, 0, B_z)$, one obtains

$$\begin{aligned} x &= x_\mathrm{c} - r\sin(s'/r + \varphi_0) \\ y &= y_\mathrm{c} + r\cos(s'/r + \varphi_0) \\ z &= z_0 + (\mathrm{d}z/\mathrm{d}s)_0 s \,. \end{aligned}$$

x_c, y_c define the position of the axis of the helix (its "centre"), s' is the projected path length, r is the radius of the projection of the helix. We have

$$\begin{aligned} x_\mathrm{c} &= x_0 + r\sin(\varphi_0) \\ y_\mathrm{c} &= y_0 - r\cos(\varphi_0) \\ \varphi_0 &= \tan^{-1}[(\mathrm{d}y/\mathrm{d}s)_0/(\mathrm{d}x/\mathrm{d}x)_0] \\ s'^2 &= s^2((\mathrm{d}x/\mathrm{d}s)_0^2 + (\mathrm{d}y/\mathrm{d}s)_0^2) \\ r &= |\boldsymbol{p}|[(\mathrm{d}x/\mathrm{d}s)_0^2 + (\mathrm{d}y/\mathrm{d}s)_0^2]^{1/2}/(q|\boldsymbol{B}|) \\ (\mathrm{d}z/\mathrm{d}s)_0^2 &= [1 - (\mathrm{d}x/\mathrm{d}s)_0^2 - (\mathrm{d}y/\mathrm{d}s)_0^2] \,. \end{aligned}$$

In the "bubble chamber convention" the dip angle λ is defined by $\sin\lambda = (\mathrm{d}z/\mathrm{d}s)$, hence

$$\cos^2\lambda = 1 - (\mathrm{d}z/\mathrm{d}s)^2 = (\mathrm{d}x/\mathrm{d}s)^2 + (\mathrm{d}y/\mathrm{d}s)^2 \,.$$

Other approximate explicit solutions for the trajectories of particles can be obtained using field symmetries allowing a simple expansion of the magnetic field, e.g. in accelerator theory. In other cases, an approximate expansion of the deviations of the field from an average value can give sufficiently precise correction formulae (e.g. in large detectors with near-homegeneous field, or in polarized targets, → [Bradamante77]). Trajectories in quadrupole fields allow a particularly elegant explicit solution (→ Quadrupole Magnet).

For numerical solutions to the equations of the trajectory in a non-homogeneous field → [Bock98] on numerical integration, Runge–Kutta methods, predictor–corrector methods, Numerov's method.

In many cases it is sufficient to know the intersection point of a particle trajectory with only few detector planes, without reference to the track behaviour elsewhere. The intersection coordinates can be expressed in terms of the initial track parameters

$$\boldsymbol{c} = \boldsymbol{f}(x_0, y_0, (\mathrm{d}x/\mathrm{d}s)_0,\ (\mathrm{d}y/\mathrm{d}s)_0,\ 1/p) \,,$$

and one can try to parameterize the function $\boldsymbol{f}$. For more details and references → [Eichinger81].

Transition Radiation. Transition radiation is produced when a relativistic particle traverses an inhomogeneous medium, in particular the boundary between materials of different electrical properties. The intensity of transition radiation is roughly proportional to the particle energy,

$$I \hat{=} m\gamma = m/\sqrt{(1-\beta^2)}\ .$$

This radiation hence offers the possibility of *particle identification* at highly relativistic energies, where Cherenkov radiation or ionization measurements no longer provide useful particle discrimination. Electron/hadron discrimination is possible for momenta from about 1 GeV/c to 100 GeV/c or higher, the upper limit being determined not only by particles reaching the Fermi plateau, but also by the radiation of highly relativistic particles.

The angular distribution of transition radiation is peaked forward with a sharp maximum at $\Theta = 1/\gamma$, hence rather collimated along the direction of the radiating particle. The total energy radiated by a single foil is found to depend on the squared difference of the plasma frequencies ω_{plasma} of the two materials; if the difference is large (e.g. $\hbar\omega_{\mathrm{air}} \approx 0.7$ and $\hbar\omega_{\mathrm{polyethylene}} \approx 20$ [eV]), the relation is

$$E \approx (2/3)\alpha\gamma\hbar\omega_{\mathrm{plasma}}\ ,$$

where $\alpha = 1/137$. The average number of radiated photons is of order $\alpha\gamma$:

$$\langle N\rangle \approx \alpha\gamma\hbar\omega_{\mathrm{plasma}}/(\hbar\langle\omega\rangle)\ .$$

The emission spectrum typically peaks between 10 and 30 keV.

In order to intensify the photon flux, periodic arrangements of a large number of foils are in use, interleaved by X-ray detectors, e.g. multiwire proportional chambers filled with xenon or a xenon/CO_2 mixture. Thin foils of lithium, polyethylene or carbon are common. Randomly spaced radiators are also in use, like foams, granules, or fibre mats.

In optimizing a detector, the ratio $n_{\mathrm{trans}}/n_{\mathrm{ion}}$ has to be maximized, where n_{trans} is the number of ions due to transition radiation entering the chamber, and n_{ion} the number of ions due to ionization inside the chamber; the latter typically results in avalanches with lower energy; however, ordinary energy loss can with some probability be confused with radiated X-rays due to the Landau tail. The optimization of the radiator/chamber sandwich has also to include the effects of attenuation of X-rays in the radiator. For details, → [Dolgoshein95], [Graham95] or [Dolgoshein93].

The signal/noise ratio can be improved by counting ionization clusters along the track ([Kleinknecht82], [Ludlam81a], [Fabjan80]). In the low-energy domain, improvements have been obtained by measuring the angle of emission of transition radiation, together with the deposited energy ([Ludlam81b], [Deutschmann81]).

For further reading, [Allison91]; for a theoretical treatment, [Ginzburg90].

Trigger. A trigger, in the context of particle detectors, is a collection of devices, usually a combination of electronics and informatics components, providing a fast signal whenever some interesting *event* has happened. Typically, a trigger is associated with some particle detector(s), and the trigger signal causes the information pertaining to these and other detector(s), or parts thereof, to be recorded or processed. The event as seen by the trigger must allow one to evaluate conditions that are predicted to be characteristic for interesting events; these conditions are often called the event *signature*. Conditions may be as simple as identifying a charged track passing through a few scintillation counters within a time gate (typical trigger in a test beam), or as complicated as effective mass criteria between identified leptons that have to be satisfied in high-energy collisions (e.g. the intended triggers at the 40 MHz Large Hadron Collider at CERN).

In many experiments, data taking, through the *dead time* it causes, is a critical factor limiting statistics and hence physics potential; an efficient trigger system is then the critical cornerstone for transmitting data that have a high probability of containing good physics, and rejecting, with respect to the possibilities of the detector, all or most of the background, viz. trivial physics or non-physics events. Clearly, not only the data elements, i.e. transmission, electronics and computing, are needed for the trigger, but equally important are detector and readout parts that provide the data to be checked in a trigger system.

Depending on the accelerator used, triggers may be gated (e.g. by bunch crossings at a collider), or permanently active (like for cosmic rays or during the flat top of a fixed-target experiment). Implementations may be synchronous and time-critical, or made from various asynchronous local subsystems operating independently in parallel, and reporting to a control unit (which also has the task of resynchronizing). The implementations of trigger conditions range from simple AND/OR gates through field-programmable gate arrays to algorithms written in general-purpose processors. Transmission delays depend on data volumes (and often on local resource occupation), and algorithms may have a data-dependent execution time; to avoid dead times at too many levels, multiple (FIFO-type) buffers smooth the fluctuations.

In large experiments, triggers are implemented in multiple levels; typically, a fast and synchronous trigger ("level 1") identifies candidate events from a subset of events, reducing the rate by some factor. Subsequently, data are digitized, transmitted to more permanent buffers and to the next (asynchronous) trigger, and more complex algorithms based on more complete data reduce the rate again ("level 2"). Eventually, after perhaps a third and fourth iteration, the entire event is transmitted to permanent storage.

Implementations of triggers not only depend on the detector design and the readout, but also on the rapidly evolving technology of data transmission and processing. The comparatively complex data flow situations can most often be understood only using Monte Carlo data and modelling programs. Queuing theory allows one to predict behaviour at the (simpler) local level. For introductory reading, → [Bock90]. Examples for trigger implementations at LEP are [Bocci95] and [Arignon93].

Trigger Efficiency. Triggers are used to bring the rate of useful events into a range manageable by the data acquisition equipment. Trigger counters also provide timing signals to the various detector parts. The trigger efficiency is mainly determined by two components:

a) The efficiency of the trigger algorithm: In the presence of a large number of topologies, fast methods to define useful event candidates will usually not lead to the required unique solution. On the other hand, useful events should not be lost, or at least not in a biased way. For a maximum available processing time a compromise must be chosen between selectivity (higher reduction of data rate) and risk of bias. A general recipe is impossible to give; useful tools for studying efficiencies of triggers are Monte Carlo methods, in combination with the toolkit of hypothesis testing (e.g. the Neyman–Pearson diagram).
b) Dead time losses: Without a higher level trigger, the frequency of recording is given by

$$f_1 = f_0/(1 + f_0\tau_r)$$

where f_1 is the frequency of recording, f_0 the raw trigger frequency, and τ_r the recording time, usually equivalent to *dead time.*

Introducing a second level trigger, i.e. a triggering algorithm that starts operating only if and when the lower-level trigger has fired, the rate of recording for good events can be improved. The relative gain is given by

$$f_2/f_1 = (1 + f_0\tau_r)/[1 + f_0(\epsilon\tau_r + \tau_f)] \,,$$

where f_2 is the rate of good events recorded with the second level trigger, ϵ is the fraction of events retained in the second level trigger, and τ_f is the decision time on the second level which may include partial data readout. In order to have any gain introduced by the second level trigger (i.e. less dead time), it is necessary that

$$f_2/f_1 > 1$$

or

$$\tau_f/\tau_r + \epsilon < 1 \,.$$

Usually, this simple algorithm is not quite applicable: the decision times for accepted and rejected events may typically have very different distributions, and some degree of parallelism is often introduced such that during the higher-level decision making, the readout starts, getting aborted if a reject decision is arrived at. Also, trigger algorithms are never fully efficient in the sense of a) above, and again compromises have to be found from case to case.

Triplicity. → Jet Variables

Two-Body Kinematics. Given two four-momenta $p = (E, \boldsymbol{p})$. There are two one-particle Lorentz invariants, the squares of the masses

$$m_1^2 = p_1^2 = E_1^2 - |\boldsymbol{p}_1|^2 \,,$$

and one two-particle invariant

$$p_1 \cdot p_2 = E_1 E_2 - \boldsymbol{p}_1 \cdot \boldsymbol{p}_2 \,.$$

From these, other Lorentz invariants can be formed such as

$$\begin{aligned}
s &= (E^{\mathrm{cm}})^2 = (p_1 + p_2)^2 = m_1^2 + m_2^2 + 2p_1 \cdot p_2 \\
Q^2 &= (p_1 - p_2)^2 = m_1^2 + m_2^2 - 2p_1 \cdot p_2 \\
F^2 &= |E_1 \boldsymbol{p}_2 - E_2 \boldsymbol{p}_1|^2 - |\boldsymbol{p}_1 \times \boldsymbol{p}_2|^2 \\
&= (p_1 \cdot p_2)^2 - m_1^2 m_2^2 \\
&= [(s - m_1^2 - m_2^2)^2/4] - m_1^2 m_2^2 \\
&= (E^{\mathrm{cm}} + m_1 + m_2)(E^{\mathrm{cm}} - m_1 - m_2)(E^{\mathrm{cm}} - m_1 + m_2) \\
&\quad (E^{\mathrm{cm}} + m_1 - m_2)/4 .
\end{aligned}$$

$E^{\mathrm{cm}} = E_1^{\mathrm{cm}} + E_2^{\mathrm{cm}}$, often denoted E^*, is the total energy in the "centre-of-mass" reference system defined by $\boldsymbol{p}_1^{\mathrm{cm}} + \boldsymbol{p}_2^{\mathrm{cm}} = 0$. E^{cm} is also the "effective mass" of the two-particle system. Q^2 is most interesting when one of the particles is incoming and the other outgoing, in a collision process. It is called the *momentum transfer* and also denoted t. F is called Möller's invariant flux factor; it is given by the area (with respect to the Minkovski metric) of the parallelogram spanned by the two four-momenta p_1 and p_2.

If particle 2 is the target particle at rest, in a collision, i.e. $\boldsymbol{p}_2 = 0$, its rest system is called the laboratory system. We have

$$F = p^{\mathrm{cm}} E^{\mathrm{cm}} = |\boldsymbol{p}_1^{\mathrm{lab}}| m_2$$

where

$$p^{\mathrm{cm}} \equiv |\boldsymbol{p}_1^{\mathrm{cm}}| = |\boldsymbol{p}_2^{\mathrm{cm}}| .$$

If the total four-momentum $P = (E, \boldsymbol{P}) = p_1 + p_2$ is given, in an arbitrary reference system, the velocity of the centre-of-mass system is

$$\boldsymbol{\beta} = \boldsymbol{P}/E = (\boldsymbol{p}_1 + \boldsymbol{p}_2)/(E_1 + E_2) .$$

Then $\boldsymbol{p}_1$ lies on an ellipsoid with principal half axes p^{cm}, p^{cm} and γp^{cm} ($\gamma \equiv 1/\sqrt{(1 - |\boldsymbol{\beta}|^2)}$), and with the centre at $E_1^{\mathrm{cm}} \boldsymbol{\beta}$, where

$$E_1^{\mathrm{cm}} = \sqrt{(m_1^2 + (p^{\mathrm{cm}})^2)} = (s + m_1^2 - m_2^2)/(2E^{\mathrm{cm}}) .$$

Define the angles Θ_1 and Θ_2 by

$$\boldsymbol{p} \cdot \boldsymbol{p}_1 = |\boldsymbol{P}||\boldsymbol{p}_1| \cos\Theta_1 .$$

The polar equation for the ellipsoid is

$$|\boldsymbol{p}_1| = \frac{(s^2 + m_1^2 - m_2^2)|\boldsymbol{P}| \cos\Theta_1 \pm 2E\sqrt{(F^2 - m_1^2|\boldsymbol{P}|^2 \sin^2\Theta_1)}}{2(s + |\boldsymbol{P}|^2 \sin^2\Theta_1)} .$$

In the special case $F = m_1|\boldsymbol{P}|$ one solution is $|\boldsymbol{p}| = 0$ and the other solution is given by

$$\begin{aligned} |\boldsymbol{p}_1| &= (s + m_1^2 - m_2^2)|\boldsymbol{P}|\cos\Theta_1/(s + |\boldsymbol{P}|^2\sin^2\Theta_1) \\ &= (2m_1E|\boldsymbol{P}|\cos\Theta_1)/(s + |\boldsymbol{P}|^2\sin^2\Theta_1) \,. \end{aligned}$$

If in addition $m_1 = m_2$, then

$$\tan\Theta_1\tan\Theta_2 = 1 - \beta^2 = 1/\gamma^2 \,.$$

This relation applies to the final state in elastic scattering for two particles with equal masses, and one particle at rest in the initial state. → also Mandelstam Variables and [Barnett96].

Two Photon Interaction Variables. → Deep Inelastic Scattering Variables.

Units. Typical for high-energy physics is the *natural unit* system, which fixes by convention two universal constants to 1:

$$\hbar = h/2\pi = 1 \qquad \text{and} \qquad c = 1 \,,$$

where h is Planck's constant and c is the speed of light in vacuum.

Thus, the number of fundamental mechanical units (L = length, M = mass, T = time) is reduced to one: Mass is usually measured in GeV in the natural unit system. The presently adopted SI (the international system of units, also known as MKSA) is related to the natural unit system through the values of three fundamental constants:

$$\begin{aligned} \hbar &= 1.054589 \cdot 10^{-34}\,\mathrm{Js}\,, \\ c &= 2.99792458 \cdot 10^{8}\,\mathrm{m\,s^{-1}}\,, \\ e &= 1.602189 \cdot 10^{-19}\,\mathrm{C} \end{aligned}$$

where e is the elementary charge.

These relations allow one to calculate the conversion factors between different units.

In spite of the attempted standardization to SI, several other unit systems are in use in electromagnetism: the Gaussian, CGS electrostatic and electromagnetic, and Heaviside–Lorentz system (for details on conversion of the units and the physical quantities → [Jackson75], appendix). Electrostatic and electromagnetic units differ only by factors of c.

The parallel use of different unit systems can produce confusion, as physical quantities are defined up to multiplicative constants, which depend on the unit system. For instance, in all unit systems the force $\boldsymbol{F}$ on a charge q in an electric field $\boldsymbol{E}$ is $\boldsymbol{F} = q\boldsymbol{E}$, hence (unit of force) = (unit of charge)·(unit of field).

However, this does not fix the units of charge and field separately, only their product. One way to fix the unit of charge is to fix by convention the proportionality constant k in Coulomb's law

$$F = kq_1q_2/r^2 \, .$$

F is the force between two point charges q_1 and q_2 separated by a distance r. In Gaussian units $k = 1$, in Heaviside–Lorentz units $k = 1/4\pi$, and in SI units

$$k = 1/4\pi\epsilon_0 = 10^{-7}\, c^2 \mathrm{kg\, m\, C^{-2}} \, ,$$

with $\varepsilon_0 = 8.854 \times 10^{-12}\ \mathrm{F\, m^{-1}}$ the permittivity of free space. Strictly speaking, the SI unit C = coulomb is defined not from Coulomb's law, but from Ampère's law for the force between parallel currents, plus the relation 1 C = 1 A · 1 s between the units for charge, current and time.

The fine structure constant is

$$\alpha = ke^2/\hbar c = 1/137.0360 \, .$$

If one combines natural mechanical units, with $\hbar = c = 1$, and Gaussian electromagnetic units, then electric charge becomes dimensionless. Thus, in these units the elementary charge is

$$e = \alpha^{1/2} = 0.08542453 \, .$$

Charge is also dimensionless in "natural Heaviside–Lorentz" units:

$$e = (4\pi\alpha)^{1/2} = 0.3028221 \, ,$$

but not in natural mechanical units combined with SI electromagnetic units.

Gaussian and Heaviside–Lorentz units are different by factors of $\sqrt{(4\pi)}$. Some conversion factors from SI units into Gaussian units (→ [Jackson75] for more detail) are:

charge q: coulomb $= 2.998 \cdot 10^9$ statcoulomb ,
electric field $\boldsymbol{E}$: volt m^{-1} $= (1/2.998) \cdot 10^{-4}$ statvolt cm^{-1} ,
magnetic induction $\bar{B}$: tesla $= 10^4$ gauss .

Hybrid unit systems are often used in which, for example, momentum is measured in GeV/c, length in m and magnetic induction ($\boldsymbol{B}$) in T = tesla. In these particular units, the elementary charge is

$$e = 0.2998 \, \mathrm{T}^{-1} \, \mathrm{m}^{-1} \, (\mathrm{GeV}/c) \, .$$

Vavilov Distribution. → Landau Distribution

Vertex Detector. A detector in collider experiments positioned as close as possible to the collision point. It is ypically made of cylindrical layers, positioned at radii of a few centimetres, the innermost layers preferrably with pixel readout. The goal of a vertex detector is to measure particle tracks very close to the interaction point (inner radii of a few cm, close to the beam pipe), thus allowing one to identify those tracks that do not come from the vertex (e.g. as a signature for short-lived decaying particles). Most vertex detectors seem to be made of semiconductor detectors, but precise drift chambers have also been used successfully, → [Abachi94].

Wavelength Shifter. Scintillating material that converts the short wavelength light ($\lambda < 400$ nm) emitted by scintillation or Cherenkov radiation (→) into a longer wavelength (blue light, $\lambda > 400$ nm) of fluorescent light, emitted isotropically.

With added wavelength shifters one obtains a considerably increased and well-defined attenuation length in plastic scintillators (→ Scintillation Counter). Light from large and/or large-surface scintillator plates can be collected into rods or plates of wavelength shifter material, a technique particularly important in large sampling calorimeters. Wavelength-shifted light also matches better the frequency sensitivity of the receiver (photomultiplier, vacuum diode). On the other hand the decay time of the signal increases to about 10 to 20 ns. In the case of a common light guide for several wavelength shifters each coupled to several scintillators, only about one sixth of the light is emitted into the cone retained by total reflection.

A good discussion is found in [Aurouet83]; → also [Bicron93].

Wilson Chamber. → Cloud Chamber

Wire Chamber. → Multiwire Chamber

References

[Aarnio91] P. Aarnio et al., The DELPHI detector at LEP, Nucl. Instrum. Meth. Phys. Res. **A303** (1991) 233.

[Aarnio95] P.A. Aarnio et al., Damage observed in silicon diodes after low energy pion irradiation, Nucl. Instrum. Meth. Phys. Res. **A360** (1995) 521.

[Abachi94] S. Abachi et al., The D0 detector, Nucl. Instrum. Meth. Phys. Res. **A338** (1994) 185.

[Abe92] F. Abe et al., Topology of three-jet events in $p\bar{p}$ collisions at $\sqrt{s} = 1.8$ TeV, Phys. Rev. **D45,5** (1992) 1448.

[Abramowicz94] H. Abramowicz, Structure of the Nucleon in Deep Inelastic Scattering, International Workshop on Deep Inelastic Scattering and Related Subjects, Eilat 1994, A. Levy, ed., World Scientific 1994.

[Abramowicz96] H. Abramowicz, Tests of QCD at low x, 28th International Conference on High Energy Physics, Z. Ajduk and A.K. Wroblewski eds., World Scientific 1996.

[Abreu96] P. Abreu et al., DELPHI Collaboration, Performance of the DELPHI detector, Nucl. Instrum. Meth. Phys. Res. **A378** (1996) 57.

[Acosto92] D. Acosto et al., Lateral shower profiles in a lead/scintillating fiber calorimeter, Nucl. Instrum. Meth. Phys. Res. **A316** (1992) 184.

[Acton93] P.D. Acton et al., A study of differences between quark and gluon jets using vertex tagging of quark jets, Z. f. Phys. **C58** (1993) 387.

[Adloff97] C. Adloff et al., Evolution of ep Fragmentation and Multiplicity in the Breit Frame, Internal Report DESY 97-108.

[Agoritsas95] V. Agoritsas et al., Scintillating fibre detectors using position-sensitive photomultipliers, Nucl. Phys. B (Proc. Suppl.) **44** (1995) 323.

[Akrawy90] M.Z. Akrawy et al., A measurement of global event shape distributions in the hadronic decays of the Z^0, Z. f. Phys. **C47** (1990) 505.

[Alekseev80] G.D. Alekseev et al., Investigation of self-quenching streamer discharge in a wire chamber, Nucl. Instrum. Meth. **177** (1980) 385.

[Allison76] W.W.M. Allison et al., The ionization loss of relativistic particles in thin gas samples and its use for particle identification (Experimental Results), Nucl. Instrum. Meth. **133** (1976) 325.

[Allison91] W.W.M. Allison and P.R.S. Wright, The Physics of Charged Particle Identification, in: Experimental Techniques in Nuclear and Particle Physics, T. Ferbel ed., World Scientific, 1991 (reprinted from: Formulae and Methods in Experimental Data Evaluation, R.K. Bock, ed., European Physical Society, 1984).

[Alner87] G.J. Alner et al., UA5: A general study of proton–antiproton physics at $\sqrt{S} = 540$ GeV, Phys. Rep. **154** (1987) 247.

[Amendolia86] S.R. Amendolia et al., Dependence of the transverse diffusion of drifting electrons on magnetic field, Nucl. Instrum. Meth. Phys. Res. **A244** (1986) 516.

[Anderhub88] H.Anderhub et al., A time expansion chamber as a vertex detector, Nucl. Instrum. Meth. Phys. Res. **A263** (1988) 1.

[Andrieu93] B. Andrieu et al., Results from pion calibration runs for the H1 calorimeter, and comparison with simulation, Nucl. Instrum. Meth. Phys. Res. **A336** (1993) 499.

[Angelini91] F. Angelini et al., The microstrip gas chamber, Nucl. Phys. B (Proc. Suppl.) **23A** (1991) 254.

[Angelescu96] T. Angelescu and A. Vasilescu, Comparative radiation hardness results obtained from various neutron sources and the NIEL problem, Nucl. Instrum. Meth. Phys. Res. **A374** (1996) 85.

[Arignon93] M. Arignon et al., The trigger system of the OPAL experiment at LEP, Nucl. Instrum. Meth. Phys. Res. **A313** (1993) 103.

[Arnault81] C. Arnault et al., A silicon aerogel counter for large acceptance hadron detection, Physica Scripta **23** (1981) 710.

[Aschenauer97] E.C. Aschenauer et al., Development of Scintillating Fiber Technology for High Rate Particle Tracking, International Europhysics Conference 1997, Jerusalem, Proceedings to be published, (preprint DESY report 97-174).

[ASTM93] ASTM E722-93: Standard practice for characterizing neutron fluence spectra in terms of an equivalent monoenergetic neutron fluence for radiation testing of electronics, American Society for Testing and Materials, Annual Book of ASTM Standards (1993).

[Aurouet83] C. Aurouet et al., Recent developments in wavelength shifters, Nucl. Instrum. Meth. Phys. Res. **211** (1983) 309.

[Barbarino79] G. C. Barbarino et al., Measurement of the second coordinate in a drift chamber using the charge division method, INFN AE 79-3 (1979).

[Barnett96] R.M. Barnett et al., Review of particle properties, Phys. Lett. **D54** (1996) 1. The URL for the *Particle Data Group* home page is http://pdg.lbl.gov/.

[Barrelet91] E. Barrelet et al., A Two-Dimensional Single Photoelectron Drift Detector for Cherenkov Ring Imaging, in: Experimental Techniques in Nuclear and Particle Physics, T.Ferbel ed., World Scientific, 1991.

[Baruzzi82] V. Baruzzi et al., Particle identification using ionization sampling in the region of the relativistic rise, Proceedings SLAC, Stanford, (1982) 109.

[Bates93] S.J. Bates, The effects of proton and neutron irradiations on silicon detectors for the LHC, Dissertation, University of Cambridge, UK, internal report RALT–006 (1993).

[Bauer96] C. Bauer et al., Recent results from the RD42 Diamond Detector Collaboration, Second International Symposium on development and Application of Semiconductor Tracking Detectors, Hiroshima 1995, Nucl. Instrum. Meth. Phys. Res. **A383** (1996) 64.

[Baur94] R. Baur et al., The CERES RICH detector system, Nucl. Instrum. Meth. Phys. Res. **A343** (1994) 97.

[Battistoni79] G. Battistoni et al., Operation of limited streamer tubes, Nucl. Instrum. Meth. **164** (1979) 57.

[Beaumont90] S.P. Beaumont et al., GaAs Detectors, Proceedings, Large Hadron Collider Workshop, Aachen (1990), CERN report 90-10, (ECFA 90-133), vol. III, 244.

[Beckers94] T. Beckers et al., Optimisation of microstrip gas chamber design and operating conditions, Nucl. Instrum. Meth. Phys. Res. **A346** (1994) 95.

[Behrend81] H.Y. Behrend et al., Cello – a new detector at PETRA, Physica Scripta **23** (1981) 610.

[Beingessner80] S.P. Beingessner and L. Bird, An extension to the standard formulae for electric fields in multiwire proportional chambers, Nucl. Instrum. Meth. **172** (1980) 613.

[Berends82] F.A. Berends et al., Radiative corrections to muon pair and quark pair production in electron-positron collisions in the Z_0 region, Nucl. Phys. **B202** (1982) 63.

[Berends83] F.A. Berends and R. Kleiss, Monte Carlo simulation of radiative corrections to the processes $e^+e^- \to \mu^+\mu^-$ and $e^+e^- \to \bar{q}q$ in the Z_0 region, Comp. Phys. Comm. **29** (1983) 185.

[Bethke91] S. Bethke, Experimental overview of jet physics and tests of QCD, Workshop on Jet Studies, Journal of Physics **G17** (1991) 1455.

[Biagi86] S. F. Biagi et al., A comparison of the charge division and timing difference techniques in streamer mode, Nucl. Instrum. Meth. Phys. Res. **A252** (1986) 586.

[Bicron93] Bicron Scintillation Products, Brochure, available from BICRON, Newbury, Ohio (USA), or P.O.B. 3093, NL-3760 DB Soest.

[Bilokou83] H. Bilokou et al., Coherent bremsstrahlung in crystals as a tool for producing high energy photon beams to be used in photoproduction experiments at the CERN SPS, Nucl. Instrum. Meth. Phys. Res. **204** (1983) 299.

[Birks64] J. Birks, Theory and Practice of Scintillation Counting, Pergamon Press, 1964.

[Bjorken63] J.D. Bjorken, Radiative corrections to inelastic electron scattering, Ann. Phys. **24** (1963) 201.

[Bjorken64] J.D. Bjorken and S.D. Drell, Relativistic Quantum Mechanics, McGraw-Hill, New York 1964.

[Blum93] W. Blum and L. Rolandi, Particle Detection with Drift Chambers, Springer, 1993.

[Bocci95] V. Bocci et al., Architecture and performance of the DELPHI trigger system, Nucl. Instrum. Meth. Phys. Res. **A362** (1995) 361.

[Bock81] R.K. Bock et al., Parameterization of the longitudinal development of hadronic showers, Nucl. Instrum. Meth. Phys. Res. **186** (1981) 533.

[Bock90] R.K. Bock, H. Grote, D. Notz, and M. Regler, Data Analysis Techniques for High-energy Physics Experiments, Cambridge University Press, 1990.

[Bock98] R.K. Bock and W. Krischer, The Data Analysis BriefBook, Springer 1998 (in print), an Internet version is available at URL http://www.cern.ch/Physics/DataAnalysis/BriefBook/).

[Bohr69] A. Bohr, B.R. Mottelson, Nuclear Structure, W.A. Benjamin 1969.

[Borders94] J.P. Borders, Optimization of Jet Energy Resolution and Response of the D0 Detector, PhD Thesis, University of Rochester, 1994.

[Bouclier70] R. Bouclier et al., Investigation of some properties of multiwire proportional chambers, Nucl. Instrum. Meth. **88** (1970) 149.

[Bouclier92] R. Bouclier et al., Development of Gas Micro-strip Chambers for High Rate Radiation Detection and Tracking, R&D project RD-28, Proposal CERN-DRDC 92-30 (1992), Status Report CERN/LHCC 96-18 (1996).

[Bourdinaud76] M. Bourdinaud et al., Use of silica aerogel for Cherenkov radiation counters, Nucl. Instrum. Meth. **136** (1976) 99.

[Bourdinaud81] M. Bourdinaud et al., Low-cost scintillators and wave length Shifters, Physica Scripta **23** (1981) 534.

[Bradamante77] F. Bradamante et al., Reconstruction of a particle moving in an axially symmetric magnetic field, Nucl. Instrum. Meth. **146** (1977) 357.

[Brandt79] S. Brandt and H.D. Dahmen, Axes and scalar measures of two-jet and three-jet events, Z. f. Phys. **C1** (1979) 61.

[Brau90] J.E. Brau, Hadron calorimetry – Optimizing performance, in: Calorimetry in High Energy Physics, International Conference at Fermilab Oct. 1990, proceedings edited by D.F. Andersen et al., World Scientific 1990.

[Brehin75] S. Brehin et al., Some observations concerning the construction of proportional chambers with thick sense wires, Nucl. Instrum. Meth. **123** (1975) 225.

[Breit36] G. Breit, E. Wigner, Capture of slow neutrons, Phys. Rev. **49** (1936) 519.

[Breit59] G. Breit, Theory of Resonance Reactions, in: Handbuch der Physik, Vol. XLI/1, Springer 1959.

[Breskin77] A. Breskin et al., High-accuracy bidimensional readout of proportional chambers with short resolution times, Nucl. Instrum. Meth. **143** (1977) 29.

[Breskin84] A. Breskin et al., On the low-pressure operation of multistep avalanche chambers, Nucl. Instrum. Meth. Phys. Res. **A220** (1984) 349.

[Breuker87] H. Breuker et al., Particle identification with the OPAL jet chamber in the region of the relativistic rise, Nucl. Instrum. Meth. Phys. Res. **A260** (1987) 329.

[Bridges81] D. Bridges et al. Initial performance of beam pipe PWC installed in CLEO at CESR, Physica Scripta **23** (1981) 655.

[Brooks79] F.D. Brooks, Development of organic scintillators, Nucl. Instrum. Meth. **162** (1979) 477.

[Bross82] A. Bross, Investigation of the Use of Charge-coupled Devices as High Resolution Position Sensitive Detectors of Ionizing Radiation, in: Proceedings, International Conference on Instrumentation for Colliding Beam Physics, SLAC, February 1982.

[Bross91] A.D. Bross and A. Pla-Dalmau, Radiation Effects in Plastic Scintillators and Fibres, International Conference on Calorimetry in HEP, Fermi National Lab, October 1990, FNAL Report 91/74.

[Brückmann88] H. Brückmann et al., On the Understanding of Sampling Calorimeters, in Conf. Proceedings: Lepton-Nucleon Interactions at High energies, February 1988, Seville, ed. by F. Barreiro, World Scientific, 1988.

[Bryant93] P.J. Bryant and K. Johnsen, The Principles of Circular Accelerators and Storage Rings, Cambridge University Press 1993.

[Burke82] D.L. Burke, Photon-Photon Collisions, Review talk, Proceedings of the 21st Int. Conf. on High Energy Physics, Paris 1982.

[Burq81] J.P. Burq et al., Observation of the Fermi plateau in the ionization energy loss of high-energy protons and pions in hydrogen gas, Nucl. Instrum. Meth. **187** (1981) 407.

[Buskulic96] D. Buskulic et al., Quark and gluon jet properties in symmetric three-jet events, Phys. Lett. **B384** (1996) 353.

[Caldwell93] A. Caldwell et al., Measurement of the time development of particle showers in a uranium scintillator calorimeter, Nucl. Instrum. Meth. Phys. Res. **A330** (1993) 389.

[Cantin74] M. Cantin et al., Silicon aerogels used as Cherenkov radiators, Nucl. Instrum. Meth. **118** (1974) 177.

[Carlson86] P. Carlson, Aerogel Cherenkov counters: construction principles and applications, Nucl. Instrum. Meth. Phys. Res. **A248** (1986) 110.

[CAS96] CERN Accelerator School publications, e.g. Frontiers of accelerator technology, World Scientific 1996. Also: Springer, Lecture Notes in Physics # 247 (1985), # 296 (1986), # 343 (1988), # 425 (1994).

[Charpak68] G. Charpak et al., The use of multiwire proportional counters to select and localize charged particles, Nucl. Instrum. Meth. **62** (1968) 262.

[Charpak73] G. Charpak, and F. Sauli, High-accuracy, two-dimensional readout in multiwire proportional chambers, Nucl. Instrum. Meth. **113** (1973) 381.

[Charpak84] G. Charpak et al., High-resolution electronic particle detectors, Ann. Rev. Nucl. Part. Sci. **34** (1984) 285.

[Chmill93] V.B. Chmill et al, An exploration of GaAs structures for solid state detectors, Nucl. Instrum. Meth. Phys. Res. **A326** (1993) 310.

[Chmill94] V.B. Chmill et al, Exploration of GaAs structures with $\pi - \nu$ junction for coordinate sensitive detectors, Nucl. Instrum. Meth. Phys. Res. **A340** (1994) 328.

[Close79] F.E. Close, An Introduction to Quarks and Partons, Academic Press 1979.

[Cobb76] J.H. Cobb et al., The ionization loss of relativistic particles in thin gas samples and its use for particle identification (theoretical predictions), Nucl. Instrum. Meth. **133** (1976) 315.

[Colas95] J. Colas, Liquid Calorimetry for LHC, in Proceedings, International Conference on Calorimetry in High-energy Physics, Upton 1994, H.A. Gordon and D. Rueger, eds., World Scientific 1995.

[Cushman92] P. Cushman, Electromagnetic and Hadronic Calorimeters, in: Instrumentation in High-energy Physics, F. Sauli, ed., World Scientific 1992.

[Dabrowski96] W. Dabrowski et al., Study of spatial resolution and efficiency of silicon strip detectors with charge division readout, Nucl. Instrum. Meth. Phys. Res. **A383** (1) (1996) 137.

[Dalitz53] R.H. Dalitz, On the analysis of τ-meson data and the nature of the τ-meson, Phil. Mag. **44** (1953) 1068.

[Damerell81] C.J.S. Damerell et al., CCD vertex detectors in particle physics, Nucl. Instrum. Meth. Phys. Res. **A342** (1994) 78.

[Damerell94] C.J.S. Damerell et al., Charge-coupled devices for particle detection with high spatial resolution Nucl. Instrum. Meth. Phys. Res. **185** (1981) 33.

[DellaMea94] G. Della Mea and F. Sauli, eds., Proceedings of the International Workshop on Microstrip Gas Chambers, Legnaro, October 1994, Edizioni Progetti, Padova 1994.

[Deutschmann81] M. Deutschmann et al., Particle identification using the angular distribution of transition radiation, Nucl. Instrum. Meth. **180** (1981) 409.

[DeWinter89] K. De Winter et al., A detector for the study of neutrino-electron scattering, Nucl. Instrum. Meth. Phys. Res. **A278** (1989) 670..

[Doke91] T. Doke, Fundamental Properties of Liquid Argon, Krypton and Xenon as Radiation Detector Media, in: Experimental Techniques in Nuclear and Particle Physics, T. Ferbel ed., World Scientific, 1991.

[Dolgoshein93] B. Dolgoshein, Transition radiation detectors, Nucl. Instrum. Meth. Phys. Res. **A326** (1993) 434.

[Dolgoshein95] B. Dolgoshein, Transition radiation trackers for the ATLAS and HERA-B experiments, Nucl. Instrum. Meth. Phys. Res. **A368** (1995) 239.

[Donaldson89] R. Donaldson, M.G.D. Gilchriese (eds.), Proceedings of the Workshop on Calorimetry for the Supercollider, Tuscaloosa, Alabama, 1989, World Scientific, 1990.

[Dorenbosch87] J. Dorenbosch et al., Calibration of the CHARM fine-grained calorimeter, Nucl. Instrum. Meth. Phys. Res. **A253** (1987) 203.

[Dreier90] P. Dreier, High resistivity silicon for detector applications, Nucl. Instrum. Meth. Phys. Res. **A288** (1990) 272.

[Dulinski83] W. Dulinski et al., A multistep avalanche chamber with charge-division read-out as a single-photon detector for ring imaging Cherenkov counters, Nucl. Instrum. Meth. Phys. Res. **217** (1983) 244.

[Durston93] S.D. Durston, A study of electromagnetic and hadronic shower shapes and position resolution, and the jet energy response of the D0 calorimeter, PhD Thesis, University of Rochester, 1993.

[Eichinger81] H. Eichinger and M. Regler, Review of track fitting methods in counter experiments, CERN Yellow Report 81-06 (1981).

[Eisele82] F. Eisele, Structure Functions, Review talk, Proceedings of the 21st Int. Conf. on High Energy Physics, Paris 1982.

[Ekelof96] T. Ekelof, ed., Proceedings of the Second International Workshop on Ring Imaging Cherenkov Detectors (RICH95) Uppsala 1995, Nucl. Instrum. Meth. Phys. Res. **A371** (1996).

[Ellis91] S.D. Ellis, Defing the undefinable: jets, workshop on jet studies, J. Phys. **G17** (1991) 1552.

[England81] J.B.A. England et al., Capacitative charge division read-out with a silicon strip detector, Nucl. Instrum. Meth. Phys. Res., **185** (1981) 43.

[Fabjan80] C.W. Fabjan and H.G. Fischer, Particle detectors, Rev. Prog. Phys. **43** (1980) 1003.

[Fabjan82] C.W. Fabjan and T.Ludlam, Calorimetry in High-Energy Physics, Ann. Rev. Nucl. Part. Sci. **32** (1982), reprinted in Formulae and Methods in Experimental Data Evaluation, R.K. Bock ed., European Physical Society 1984.

[Fabjan91] C.W. Fabjan, Calorimetry in High-energy Physics, in: Experimental Techniques in Nuclear and Particle Physics, T. Ferbel ed., World Scientific, 1991.

[Fabjan95a] C.W. Fabjan, Liquid ionization calorimetry: review and preview, Nucl. Instrum. Meth. Phys. Res. **A360** (1995).

[Fabjan95b] C.W. Fabjan et al., The TIC – a multi-particle threshold imaging Cherenkov detector, Nucl. Instrum. Meth. Phys. Res. **A367** (1995) 240.

[Fanet91] H. Fanet and J.C. Lugol, Measurement of charge with an active integrator in the presence of noise and pile-up effects, Nucl. Instrum. Meth. Phys. Res. **A301** (1991) 295.

[Fano63] U. Fano, Penetration of protons, alpha particles and mesons, Ann. Rev. Nucl. Part. Sci. **13** (1963) 1.

[Fassò93] A. Fassò et al., in: Calorimetry in High Energy Physics, International Conference at La Biodola, proceedings edited by A. Menzione and A. Scribano, World Scientific 1993.

[Fenker91] H. Fenker et al., Progress in the Use of Avalanche Photodiodes for Readout of Calorimeters, Report SSCL-552, Superconducting Collider Laboratory, 1991.

[Fermi51] E. Fermi, Nuclear Physics, Lecture Notes, Chicago University Press, Rev. Edition 1959.

[Fernow86] R.C. Fernow, Introduction to Experimental Particle Physics, Cambridge University Press, 1986.

[Ferrari93] A. Ferrari and P.R. Sala, A New Model for Hadronic Interactions at intermediate Energies, in: Proceedings of the MC93 Conference, Tallahassee 1993, eds. P. Dragovitsch et al., World Scientific 1993.

[Ferrari97] A. Ferrari and P.R. Sala, Intermediate and High Energy Models in FLUKA, in: Proceedings of the International Conference on Nuclear Data for Science and Technology, (in press) Trieste 1997.

[Ferrere92] D. Ferrere et al., High resolution crystal calorimetry at LHC, Nucl. Instrum. Meth. Phys. Res. **A315** (1992) 332.

[Fesefeldt85] H. Fesefeldt, The Simulation of Hadronic Showers, Physics and Applications, RWTH Aachen Report PITHA 85/02.

[Fesefeldt90a] H. Fesefeldt, T.Hamacher, and J.Schug, Tests of punch-through simulation, Nucl. Instrum. Meth. Phys. Res. **A292** (1990) 279.

[Fesefeldt90b] H. Fesefeldt, GEANT 3.14 Benchmark Tests, in: Calorimetry in High Energy Physics, International Conference at Fermilab Oct. 1990, proceedings edited by D.F. Andersen et al., World Scientific 1990.

[Feynman69] R.P. Feynman, Very high energy collisions of hadrons, Phys. Rev. Lett. **23** (1969) 1415.

[Fretwurst96] E. Fretwurst et al., An investigation into the radiation damage of the silicon detectors of the H1-PLUG calorimeter, Nucl. Instrum. Meth. Phys. Res. **A372** (1996) 368.

[Fricke86] J. Fricke, Aerogels, Springer 1986.

[Friedländer91] E.M. Friedländer et al., Multiplicity distributions, transverse momenta, and nonstationary effects, Phys. Rev. **D44,5** (1991) 1396.

[Gary94] G.W. Gary, Multiplicity difference between quark and gluon jets, Phys. Rev. **D49,9** (1994) 4503.

[Gatti79] E. Gatti et al., Optimum geometry for strip cathodes or grids, Nucl. Instrum. Meth. **163** (1979) 83.

[Gatti81] E. Gatti et al., Analysis of the position resolution in centroid measurements in MWPC, Nucl. Instrum. Meth. Phys. Res. **188** (1981) 327.

[Gatti84] E. Gatti and P. Rehak, Semiconductor drift chambers – an application of a novel charge transport scheme, Nucl. Instrum. Meth. Phys. Res. **225** (1984) 608.

[Gatti88] E. Gatti et al., Silicon drift chamber prototype for the upgrade of the UA6 experiment, Nucl. Instrum. Meth. Phys. Res. **A273** (1988) 865.

[Geijsberts92] M. Geijsberts et al., Tests of the performance of different gas mixtures in microstrip gas counters, Nucl. Instrum. Meth. Phys. Res. **A313** (1992) 377.

[Gerdt80] V.P. Gerdt et al., Analytic calculations on digital computers for applications in physics and mathematics, Sov. Phys. Usp. **23** (1980) 59.

[Ginneken89] A. van Ginneken, Non ionizing energy deposition in silicon for radiation damage studies, Fermi Nat. Acc. Lab. report FN-522 (1989).

[Ginzburg90] V.L. Ginzburg and V.N. Tsytovich, Transition Radiation and Transition Scattering, Adam Hilger 1990.

[Gluckstern63] R.L. Gluckstern, Uncertainties in track momentum and direction due to multiple scattering and measurement errors, Nucl. Instrum. Meth. **24** (1963) 381.

[Gordon95] H.A. Gordon and D. Rueger, eds., Proceedings of the International Conference on Calorimetry in High-energy Physics, Brookhaven Sept 1994, World Scientific 1995.

[Graf90] N.A. Graf, The D0 Shower Library, in: Calorimetry in High Energy Physics, International Conference at Fermilab Oct. 1990, proceedings edited by D.F. Andersen et al., World Scientific 1990.

[Graham95] G.E. Graham et al., Design and test results of a transition radiation detector for a Fermilab fixed target rare kaon decay experiment, Nucl. Instrum. Meth. Phys. Res. **A367** (1995) 224.

[Gratta94] G. Gratta, H.Newman, and R.Y.Zhu, Crystal calorimeters in particle physics, Ann. Rev. Nucl. Part. Sci. **44** (1994) 453.

[Hall90] G. Hall, Prospects for silicon detectors in the 1990s, 4th Topical Seminar on Experimental Apparatus for High-energy Physics and Astrophysics, San Miniato, 1990.

[Hallewell96] G.D. Hallewell, The present status of pixel detector development for high energy physics collider applications, Second International Symposium on Development and Application of Semiconductor Tracking Detectors, Hiroshima 1995, Nucl. Instrum. Meth. Phys. Res. **A383** (1996) 44.

[Hamilton72] J. Hamilton and B. Tromborg, Partial Wave Amplitudes and Resonance Poles, Oxford Clarendon Press, 1972.

[Harigel94] G.G. Harigel, D.C. Colley, and D.C. Cundy, eds., Proceedings of the Conference on the Bubble Chamber and its Contributions to Particle Physics, Geneva, Nucl. Phys. B (Proc. Suppl.) **36** (1994).

[Heath79] R.L. Heath et al., Inorganic scintillators, Nucl. Instrum. Meth. **162** (1979) 431.

[Heijne80] E.H.M. Heijne et al., A Silicon surface barrier microstrip detector designed for high energy physics, Nucl. Instrum. Meth. **178** (1980) 331.

[Heijne83] E.H.M. Heijne, Muon Flux Measurement with Silicon Detectors in Neutrino Beams, CERN Yellow Report 83-06 (1983).

[Heijne89] E.H.M. Heijne and P. Jarron, Silicon Detector Development in Europe, in: 1988 Snowmass Summer Study on High-energy Physics in the 1990-s, p. 739, World Scientific 1989.

[Heintze78] J. Heintze, Drift chambers and recent developments, Nucl. Instrum. Meth. **156** (1978) 227.

[Henning81] S. Henning et al., Production of silicon aerogel, Physica Scripta **23** (1981) 697.

[Hertzog90] D. Hertzog et al., A high-resolution lead/scintillating fibre electromagnetic calorimeter, Nucl. Instrum. Meth. Phys. Res. **A294** (1990) 446.

[Holder78] H. Holder et al., Performance of a magnetized total absorption calorimeter between 15 GeV and 140 GeV, Nucl. Instrum. Meth. **151** (1978) 69.

[Holl96] P. Holl et al., New developments in radiation detectors, Conference Proceedings, Elmau, 1995, Nucl. Instrum. Meth. Phys. Res. **A377** (1996) 346.

[Hooft73] G. 't Hooft and M. Veltman, Diagrammar, CERN Yellow Report 73-09 (1973).

[Huhtinen93a] M. Huhtinen and P.A. Aarnio, Hadron fluxes in inner parts of LHC, Nucl. Instrum. Meth. Phys. Res. **A336** (1993) 98.

[Huhtinen93b] M. Huhtinen and P. Aarnio, Estimation of pion induced damage in silicon, SEFT report HU–SEFT R 1993-02, Helsinki 1993.

[Hyams83] B. Hyams et al., A Silicon counter telescope to study short-lived particles in high-energy hadronic interactions, Nucl. Instrum. Meth. Phys. Res. **205** (1983) 99.

[Iaselli92] G. Iaselli et al., A statistical hadron shower parameterization for sampling calorimeters, Nucl. Instrum. Meth. Phys. Res. **A311** (1992) 122.

[ICRUReport33] International Commission on Radiation Units and Measurements (ICRU) Report 33 – Radiation quantities and units (1980).

[INTE94] International Symposium on Aerogels, Berkeley, Sept. 1994, Journal of Non-Crystalline Solids, **185,6** (1995). There is also a web site (Microstructured Materials Group of LBL):
http://eande.lbl.gov/ECS/aerogels/aerogels.htm

[Jackson75] J.D. Jackson, Classical Electrodynamics, Wiley, New York, 1975.

[Jadach97] S. Jadach et al., eds., Proceedings, 3rd International Symposium on Radiative Corrections (CRAD '96), Acta Phys. Pol. **B28** (3-4) (1997).

[James83] F. James, Fitting tracks in wire chambers using the Chebyshev norm instead of least squares, Nucl. Instrum. Meth. Phys. Res. **211** (1983) 145.

[Jauch80] J.M. Jauch, F. Rohrlich, The Theory of Photons and Electrons, Springer 1980.

[Jonker83] M. Jonker et al., The limited streamer tube system of the CHARM Collaboration, Nucl. Instrum. Meth. Phys. Res. **215** (1983) 361.

[Kadyk91] J.A. Kadyk, Wire chamber aging, Nucl. Instrum. Meth. **A300** (1991) 436.

[Kazovsky96] L. Kazovsky, S. Benedetto, and A. Willner, Optical Fiber Communication Systems, Artech House, 1996.

[Kemmer80] J. Kemmer, Fabrication of low noise silicon radiation detectors by the planar process, Nucl. Instrum. Meth. **169** (1980) 499.

[Klanner85] R. Klanner, Silicon detectors, Nucl. Instrum. Meth. Phys. Res. **A235** (1985) 209.

[Kleinknecht82] K. Kleinknecht, Particle detectors, Physics Reports **84,2** (1982).

[Koba72] Z. Koba et al., Scaling of multiplicity distributions in high energy hadron collisions, Nucl. Phys. **B40** (1972) 317.

[Kölbig83] K.S. Kölbig and B. Schorr, A program package for the Landau distribution, Comp. Phys. Comm. **31** (1984) 97.

[Konobeyev92] A.Yu Konobeyev et al., Neutron displacement cross-sections for structural materials below 800 MeV, J. Nucl. Mater. **186** (1992) 117.

[Krammer95] M. Krammer et al., eds., Proceedings, Wire Chamber Conference, Vienna, Nucl. Instrum. Meth. Phys. Res. **A367** (1995).

[Landau44] L. Landau, J. Physics (USSR) **8** (1944) 201. Also: Collected Papers, D. ter Haar ed., Pergamon Press, Oxford, 1965.

[Landua96] R. Landua, New Results in Spectroscopy, 28th International Conference on High Energy Physics, Z. Ajduk and A.K. Wroblewski, eds., World Scientific 1996.

[Lazanu97] I. Lazanu et al., Non-ionising energy loss of pions in thin silicon samples, Nucl. Instrum. Meth. Phys. Res. **A388** (1997) 370.

[Lazo86] M.S. Lazo et al., Silicon and silicon dioxide neutron damage functions, Sandia Nat. Lab. Tech. rep. SAND 87–0098, vol.1, Proc. Fast Burst React. Workshop 1986, 85

[Lecoq92] P. Lecoq, Results on New Scintillating Crystals from the Crystal Clear Collaboration, IEEE Nucl. Science Symposium, Orlando 1992.

[Lecoq93] P. Lecoq et al., The Quest for New Scintillators, Proceedings of Calorimetry in High-energy Physics, La Biodola 1993, A.Menzione and A. Scribano eds., World Scientific, 1993.

[Lehraus83] I. Lehraus, Progress in particle identification by ionization sampling, Nucl. Instrum. Meth. Phys. Res. **217** (1983) 43.

[Leo94] W.R. Leo, Techniques for Nuclear and Particle Physics Experiments, Springer, 1994.

[Leroy86] C. Leroy, Y. Sirois, and R. Wigmans, An experimental study of nuclear fission to the signal of uranium hadron calorimeters, Nucl. Instrum. Meth. Phys. Res. **A252** (1986) 4.

[Lint87] V.A.J. van Lint, The physics of radiation damage in particle detectors, Nucl. Instrum. Meth. **A253** (1987) 453.

[Litchfield84] P. Litchfield, Partial Wave Analysis, in: Formulae and Methods in Experimental Data Evaluation, R.K.Bock, ed., European Physical Society, 1984

[Livan95] M. Livan et al., Scintillating-Fibre calorimetry, CERN Yellow Report 95-02 (1995).

[Livingston37] M.S. Livingston and H.A. Bethe, Nuclear dynamics, experimental, Rev. Mod. Phys. **9** (1937) 285.

[Lohrmann81] E. Lohrmann, Hochenergiephysik, Teubner Studienbücher, 1981.

[Lohse92] T. Lohse and W. Witzeling, The Time Projection Chamber, in: Instrumentation in High Energy Physics, F. Sauli, ed., World Scientific 1992.

[Longo75] E. Longo et al., Monte Carlo Calculations of Photon-initiated Electromagnetic Showers in Lead Glass, Nucl. Instrum. Meth. **128** (1975) 283.

[Ludlam81a] T. Ludlam et al., Particle Identification by Electron Cluster Detection of Transition Radiation, Nucl. Instrum. Meth. **180** (1981) 413.

[Ludlam81b] T. Ludlam et al., Particle Identification using the Angular Distribution of Transition Radiation, Nucl. Instrum. Meth. **180** (1981) 409.

[Lutz95] G. Lutz and A.S. Schwarz, Silicon devices for charged-particle track and vertex detection, Ann. Rev. Nucl. Part. Sci. **45** (1995) 295.

[Majewski92] S. Majewski, Crystal Scintillators, in: Instrumentation in High-energy Physics, F. Sauli, ed., World Scientific 1992.

[Mandelstam58] S. Mandelstam, Determination of the Pion-Nucleon Scattering Amplitude from Dispersion Relations and Unitarity, Phys. Rev. **112** (1958) 1344.

[Marini85] G. Marini et al., Radiation Damage to Organic Scintillation Materials, CERN internal report 85-08.

[Martin93] A.D. Martin et al., Parton distributions updated, Phys. Lett. **B306** 1993) 145.

[Martin94] A.D. Martin, Structure Functions and small x Physics, International Workshop on Deep Inelastic Scattering and Related Subjects, Eilat 1994, A. Levy, ed., World Scientific 1994.

[Marx78] J.N. Marx and D.R. Nygren, The time projection chamber, Physics Today, October 1978 p. 46.

[Matthews81] J.L. Matthews et al., The distribution of electron energy losses in thin absorbers, Nucl. Instrum. Meth. **180** (1981) 573.

[Maximon69] I.C. Maximon, Comments on radiative corrections, Rev. Mod. Phys. **41** (1969) 193.

[McKemey96] A. McKemey, The SLC CCD pixel vertex detectors and its upgrade, Nuovo Cimento **109A,6-7** (1996) 1027.

[Mo69] L.W. Mo et al., Radiative corrections to elastic and inelastic ep and μp scattering, Rev. Mod. Phys. **41** (1969) 205.

[Montanet83] F. Montanet, Utilisation d'un Compteur Cherenkov a Radiateur d'Aerogel de Silice, Internal Report LAPP (IN2P3) 1983.

[Moyal55] J.E. Moyal, Theory of ionization fluctuations, Phil. Mag. **46** (1955) 263.

[Nelson85] W.R. Nelson, H. Hirayama, and D.W.O. Rogers, The EGS4 Code System, SLAC Report 265 (1985).

[Nonaka96] N. Nonaka et al., An APD linear array for scintillating fibre tracker readout, Second International Symposium on Development and Application of Semiconductor Tracking Detectors, Hiroshima 1995, Nucl. Instrum. Meth. Phys. Res. **A383** (1996) 81.

[O'Brien91] E. O'Brien et al., A Transition Radiation Detector for RHIC featuring accurate tracking and $\mathrm{d}E/\mathrm{d}x$ Particle Identification, in: Y. Makdisi and A.J. Stevens, eds., Proc. Symposium on RHIC Detector Research and Development, Brookhaven National Laboratory, BNL 52321 (1991).

[Oed88] A. Oed, Position-sensitive detector with microstrip anode for electron multiplication with gases, Nucl. Instrum. Meth. Phys. Res. **A263** (1988) 351.

[Ougouag90] A.M. Ougouag et al., Differential displacement KERMA cross sections for neutron interactions in Si and GaAs, IEEE Trans. Nucl. Sci. **NS-37,6** (1990) 2219.

[Paul69] E.B. Paul, Nuclear and Particle Physics, North Holland 1969.

[Peisert84] A. Peisert and F. Sauli, Drift and Diffusion of Electrons in Gases: a Compilation, CERN Report 84-08, 13 July 1984.

[Peisert92] A. Peisert, Silicon Microstrip Detectors, in: Instrumentation in High Energy Physics, F. Sauli, ed., World Scientific 1992.

[Pentia96] M. Pentia et al., A fast procedure for geometrical parameter determination of a silicon vertex tracker, Nucl. Instrum. Meth. Phys. Res. **A369** (1996) 101.

[Piuz82] F. Piuz et al., Evaluation of systematic errors in the avalanche localization along the wire with cathode strips read-out MWPC, Nucl. Instrum. Meth. Phys. Res. **196** (1982) 451.

[Piuz83] F. Piuz, Measurement of the longitudinal diffusion of a single electron in gas mixtures used in proportional counters, Nucl. Instrum. Meth. Phys. Res. **205** (1983) 425.

[Poelz86] G. Poelz, Aerogel Cherenkov counters at DESY, Nucl. Instrum. Meth. Phys. Res. **A248** (1986) 118.

[Radeka80] V.Radeka, R.A. Boie, Centroid finding method for position-sensitive detectors, IEEE Trans. Nucl. Sci. **NS27** (1980) 351.

[Radeka91] V.Radeka, Signal and Noise in Position-sensitive Detectors, in: Experimental Techniques in Nuclear and Particle Physics, T. Ferbel ed., World Scientific, 1991

[Richard71] C. Richard-Serre, Evaluation de la Perte d'Energie Unitaire et du Parcours pour des Muons de 2 à 600 GeV dans un Absorbant quelconque, CERN Yellow Report 71-18 (1971).

[Rossi65] B. Rossi, High-Energy Particles, Prentice Hall Series, 1965.

[Rushbrooke81] J.G. Rushbrooke, The UA5 Streamer Chamber Experiment at the SCS $p\bar{p}$ Collider, Phys. Scripta **23** (1981) 642.

[Sadoulet82] B. Sadoulet, Limits on the Accuracy of Drift Chambers and Calibration Beams, in: Proceedings, International Conference on Instrumentation for Colliding Beam Physics, SLAC, February 1982.

[Sangster56] R.C. Sangster and J.W. Irwine, Study of organic scintillators, Journ. Chem. Phys. **24** (1956) 670.

[Sauli83] F. Sauli, New Developments in Gaseous Detectors, Techniques and Concepts in High Energy Physics 2, NATO Advanced Institute, Lake George, Plenum 1983.

[Sauli91] F.Sauli, Principles of Operation of Multiwire Proportional and Drift Chambers, in: Experimental Techniques in Nuclear and Particle Physics, T.Ferbel ed., World Scientific, 1991.

[Scharf86] W. Scharf, Particle Accelerators and their Uses, Harwood, 1986.

[Schmitz94] J. Schmitz, The Microstrip Gas Counter and its application in the ATLAS inner tracker, PhD Thesis, University of Amsterdam and NIKHEF, 1994.

[Schorr74] B. Schorr, Programs for the Landau and the Vavilov distributions and the corresponding random numbers, Comp. Phys. Comm. **7** (1974) 215.

[Schultz77] G. Schultz, G. Charpak and F. Sauli, Mobility of positive ions in some gas mixtures, Rev. Phys. Appl. **12** (1977) 67.

[Scott63] W.T. Scott, The theory of small-angle scattering of fast charged particles, Rev. Mod. Phys. **35** (1963) 231.

[Serre67] C. Serre, Evaluation de la Perte d'Energie Unitaire et du Parcours, CERN Yellow Report 67-5 (1967).

[Shutt67] R.P. Shutt, ed., Bubble and Spark Chambers, Academic Press, 1967.

[Sill90a] A. F. Sill, Advanced field shaping drift chambers for SSC muon tracking, Proceedings, Symposium on Detector Research and Development for the SSC, Fort Worth, T. Dombeck et al., eds., World Scientific (1991) 225.

[Sill90b] A. F. Sill, Advanced field shaping drift chambers for SSC forward tracking, Proceedings, Symposium on Detector Research and Development for the SSC, Fort Worth, T. Dombeck et al., eds., World Scientific (1991) 655.

[Sipilä80] H. Sipilä et al., Mathematical treatment of space charge effects in proportional counters, Nucl. Instrum. Meth. **176** (1980) 381.

[Smith96] K.M. Smith, GaAs detector status, Second International Symposium on Development and Application of Semiconductor Tracking Detectors, Hiroshima 1995, Nucl. Instrum. Meth. Phys. Res. **A383** (1996) 75.

[Sternheimer71] R.M. Sternheimer and R.F. Peierls, General expression for the density effect for the ionization loss of charged particles, Phys. Rev. **B3** (1971) 3681.

[Stoks93] V. G. J. Stoks et al., Partial wave analysis of all nucleon-nucleon scattering data below 350 MeV, Phys. Rev. **C48,2** (1993) 792.

[Summers93] G.P. Summers et al., Damage correlations in semiconductors exposed to gamma, electron and photon radiation, IEEE Trans. Nucl. Sci. **NS-40,6** (1993) 1372.

[Sze81] S. M. Sze, The Physics of Semiconductor Devices, 2nd edition, Wiley Interscience (1981).

[Teitelbaum92] L.T. Teitelbaum, Charged Particle Spectra in ${}_{32}S + {}_{32}S$ Interactions at 200 GeV/nucleon from CCD-imaged Nuclear Collisions in a Streamer Chamber, PhD Thesis, Lawrence Berkeley Laboratory, LBL Report 32812, April 1992.

[Treille96] D. Treille, The physics potential of the RICH, Nucl. Instrum. Meth. Phys. Res. **A371** (1996) 178.

[Tsukamoto96] T. Tsukamoto et al., Properties of a CCD sensor for vertex detector applications, Second International Symposium on development and Application of Semiconductor Tracking Detectors, Hiroshima 1995, Nucl. Instrum. Meth. Phys. Res. **A383** (1996) 256.

[Vacchi93] A. Vacchi et al., Beam tests of a large area silicon drift detector, Nucl. Instrum. Meth. Phys. Res. **A326** (1993) 267.

[Varelas96] N. Varelas, Jet Studies at all Rapidities from D0 and CDF, Proceedings of the 31st Rencontres de Morions, March 1996, Editions Frontieres.

[Vasilescu97] A. Vasilescu, The NIEL scaling hypothesis applied to neutron spectra of irradiation facilities and in the ATLAS and CMS SCT, internal report ROSE/TN/97–2.

[Va'vra92] J. Va'vra, Wire chamber gases, Nucl. Instrum. Meth. Phys. Res **A323** (1992) 34.

[Villa86] F. Villa, ed., Vertex detectors, Workshop proceedings, Erice, Plenum Press (1988).

[Walenta71] A.H. Walenta et al., The multiwire drift chamber: A new type of proportional wire chamber, Nucl. Instrum. Meth. **92** (1971) 373.

[Walenta79a] A.H. Walenta et al., Measurement of the ionization loss in the region of relativistic rise for noble and molecular gases, Nucl. Instrum. Meth. **161** (1979) 45.

[Walenta79b] A.H. Walenta, The time expansion chamber and single ionization cluster measurement, IEEE Trans. Nucl. Sci. **NS - 26.1** (1979) 73.

[Walenta82] A.H. Walenta et al., The time expansion chamber as a high precision drift chamber, in: Proceedings, International Conference on Instrumentation for Colliding Beam Physics, SLAC, February 1982.

[Walraff91] W. Walraff, Prospects for Large Scintillating Noble Liquid Calorimeters, in: Proceedings of Calorimetry in High-energy Physics, Capri 1991, A.Ereditato ed., World Scientific, 1991.

[Ward95] B.F.L. Ward, ed., International Symposium on Radiative Corrections : Status and Outlook, Proceedings, Gatlinburg 1994, World Scientific, 1995.

[Webber95] B.R. Webber, Hadronic Final States, in: Workshop on Deep Inelastic Scattering and QCD, J.F. Laporte and Y. Sirois, eds., Paris 1995.

[Weber95] M.J. Weber, Scintillator Materials for Calorimetry, in Proceedings of the International Conference on Calorimetry in High-energy Physics, Upton 1994, H.A.Gordon and D.Rueger, eds., World Scientific 1995.

[Wigmans87] R.Wigmans, On the energy resolution of uranium and other hadron calorimeters, Nucl. Instrum. Meth. **A259** (1987) 389.

[Wigmans91a] R.Wigmans, Advances in hadron calorimetry, Ann. Rev. Nucl. Part. Sci. **41** (1991).

[Wigmans91b] R.Wigmans, High-resolution Hadron Calorimetry, in: Experimental Techniques in Nuclear and Particle Physics, T.Ferbel ed., World Scientific, 1991.

[Wu80] Sau Lan Wu, A simple Method of Four-jet Analysis in e^+e^- Annihilation, DESY Report 80/127.

[Ypsilantis81] T. Ypsilantis, Cherenkov ring imaging, Physica Scripta **23** (1981) 371.

[Ypsilantis94] T. Ypsilantis and J. Seguinot, Theory of ring imaging Cherenkov counters, Nucl. Instrum. Meth. Phys. Res. **A343** (1994) 30.

[Zhu90] Ren-yuan Zhu, Limits to the Precision of Electromagnetic Calorimeters, in: Calorimetry in High Energy Physics, International Conference at Fermilab Oct. 1990, proceedings edited by D.F. Andersen et al., World Scientific 1990.

[Zorn92] C. Zorn, Organic Scintillators, in: Instrumentation in High-energy Physics, F. Sauli, ed., World Scientific 1992.

Springer and the environment

At Springer we firmly believe that an international science publisher has a special obligation to the environment, and our corporate policies consistently reflect this conviction.
We also expect our business partners – paper mills, printers, packaging manufacturers, etc. – to commit themselves to using materials and production processes that do not harm the environment. The paper in this book is made from low- or no-chlorine pulp and is acid free, in conformance with international standards for paper permanency.

Druck: Strauss Offsetdruck, Mörlenbach
Verarbeitung: Schäffer, Grünstadt